ADVANCES IN INDUSTRIAL MANUFACTURING AND MATERIALS

Integrating a New Era of Computational Modeling and Machine Learning

ADVANCES IN INDUSTRIAL MANUFACTURING AND MATERIALS

Integrating a New Era of Computational Modeling and Machine Learning

Debesh Mishra, PhD
Swagatika Mishra, PhD
Editors

AAP | APPLE ACADEMIC PRESS

First edition published 2026

Apple Academic Press Inc.
1265 Goldenrod Circle, NE,
Palm Bay, FL 32905 USA

760 Laurentian Drive, Unit 19,
Burlington, ON L7N 0A4, Canada

CRC Press
2385 NW Executive Center Drive,
Suite 320, Boca Raton FL 33431

4 Park Square, Milton Park,
Abingdon, Oxon, OX14 4RN, UK

Apple Academic Press exclusively co-publishes with CRC Press, an imprint of Taylor & Francis Group, LLC

Library and Archives Canada Cataloguing in Publication

CIP data on file with Canada Library and Archives

Library of Congress Cataloging-in-Publication Data

CIP data on file with US Library of Congress

ISBN: 978-1-77964-046-8 (hbk)
ISBN: 978-1-77964-047-5 (ebk)

DOI: 10.1201/9781779640475

About the Editors

Debesh Mishra, PhD
Associate Professor, Mechanical Engineering,
IES University, Bhopal, Madhya Pradesh, India

Debesh Mishra, PhD, is presently working as an Associate Professor and Head of the Department of Mechanical Engineering at IES University, Bhopal, India. He has more than 12 years of teaching experience. In addition, he has published more than 40 articles in reputed journals and international conferences. He has also contributed 18 book chapters and co-authored four internationally recognized books. His areas of interest include ergonomics, occupational health and safety, sustainability, supply chain management, and optimization techniques. Dr. Mishra obtained his PhD in Mechanical Engineering from KIIT Deemed to be University, Odisha, India. He completed his MTech with a specialization in Industrial Engineering and Management, and his BE in Mechanical Engineering.

Swagatika Mishra, PhD
Assistant Professor, Mechanical Engineering,
VSSUT, Burla, Sambalpur, Odisha, India

Swagatika Mishra, PhD, is an Assistant Professor in the Department of Mechanical Engineering at Veer Surendra Sai University of Technology (VSSUT), Burla, Odisha—a UGC-recognized state government university. She has published more than 30 articles in various international journals and conferences. She has authored several book chapters on nontraditional machining and supply chain management. In addition, she has served as a reviewer for many reputed journals and international conferences. One seed fund research project under OURIIP is currently underway under her supervision. Her areas of interest include additive manufacturing, nontraditional machining, production and operations management, operations research, and supply chain management. With more than 13 years of teaching and research experience, she has guided many MTech and BTech students.

Contents

Contents

Contributors

Vijay Kumar Ambati
Department of Mechanical Engineering, Raghu Engineering College (A), Dakamarri, Visakhapatnam, Andhra Pradesh, India

S. Beura
Department of Mechanical Engineering, ITER, SOA (Deemed to be) University, Bhubaneswar, Odisha, India

Bibhu Prasad Dash
Department of Textile Engineering, Odisha University of Technology and Research, Bhubaneswar, Odisha, India

Mavuso Ayanda Faith
Durban University of Technology, Durban, KwaZulu-Natal, South Africa

Hafsa Jamshaid
Faculty of Textile Engineering, National Textile University, Faisalabad, Punjab, Pakistan

John Kamit
Papua New Guinea University of Technology, Lae, Morobe Province, Papua New Guinea

D. S. Chaitanya Kishore
Department of Mechanical Engineering, JNTU Anantapur, Anantapur, Andhra Pradesh, India

Amit Kumar Mehar
Department of Mechanical Engineering, Raghu Engineering College (A), Dakamarri, Visakhapatnam, Andhra Pradesh, India

Rajesh Mishra
Department of Material Science and Manufacturing Technology, Faculty of Engineering, Czech University of Life Sciences Prague, Prague-Suchdol, Prague, Czech Republic

U. K. Mohanty
Department of Mechanical Engineering, ITER, SOA (Deemed to be) University, Bhubaneswar, Odisha, India; Department of Metallurgical and Materials Engineering, NIT Rourkela, Odisha, India

Guambo Mondo
Papua New Guinea University of Technology, Lae, Morobe Province, Papua New Guinea

Kamalakanta Muduli
Papua New Guinea University of Technology, Lae, Morobe Province, Papua New Guinea

Ramesh Naidu
Department of Mechanical Engineering, JNTU Anantapur, Anantapur, Andhra Pradesh, India

S. N. Pati
Department of Mechanical Engineering, ITER, SOA (Deemed to be) University, Bhubaneswar, Odisha, India

Srinivasu Rokkala
Department of Mechanical Engineering, Raghu Engineering College (A), Dakamarri, Visakhapatnam, Andhra Pradesh, India

Arun Kumar Rout
Department of Mechanical Engineering, OP Jindal University, Raigarh, Chhattisgarh, India

P. K. Satapathy
Department of Mechanical Engineering, ITER, SOA (Deemed to be) University, Bhubaneswar, Odisha, India

Anuj Shukla
Durban University of Technology, Durban, KwaZulu-Natal, South Africa

Jitesh Kumar Singh
Department of Production Engineering, Veer Surendra Sai University of Technology, Burla, Odisha, India

Abbreviations

γ-APS	γ-aminopropyltriethoxysilane
BF	blast furnace
CIE	Commission Internationale de l'Éclairage
CIPET	Central Institute of Plastics Engineering and Technology
CM	condition monitoring
CNT	carbon nanotube
CPCB	central pollution control board
CTDI	computed tomography dose index
CTE	coefficients of thermal expansion
DT	decision tree
EBCS	electronic bonus-card system
ET	extra trees
FEA	finite-element analysis
FEM	finite-element method
FICCI	Federation of Indian Chambers of Commerce and Industry
FPs	fabric parameters
FSP/FSPed	friction stir processing
FSW	friction stir welding
Gr	graphite
HCHC	high-carbon, high-chromium
HE	high elongation
HGBM	histogram-based gradient boosting machine
HMI	human-machine interface
IDF	inverse document frequency
IFSS	interfacial shear stress
IoT	internet of things
LE	low elongation
LGBM	light gradient boost machine
LSVM	linear support vector machine
MSWM	municipal solid waste management
MWCNT	multi-walled carbon nanotube
NC	nanocomposite
NEERI	National Engineering and Environmental Research Institute
NFRC	natural fiber-reinforced composites
NH4	ammonium group
PCP	percentage categorical pruned
PDEs	partial differential equations

PdM/PM	predictive maintenance
PvM	preventive maintenance
RF	random forest
RMSE	root-mean-squared error
SCE	standard calomel electrode
SFRCs	synthetic fiber-reinforced composites
SiC	silicon carbide
SPD	spectral power distribution
SVM	support vector machine
SWOT	strengths, weaknesses, opportunities, and threats
VIDS	virtual image display system
VWS	visual wear system
WC	tungsten carbide
WEEE	waste of electrical and electronic equipment
WHO	World Health Organization
XAI	explainable artificial intelligence
XB	XGBoost

Acknowledgments

I would like to take this opportunity to acknowledge the many people who have contributed, directly or indirectly, to the compilation of this book. I am deeply thankful and grateful to all those who have made this book possible. Furthermore, my sincere gratitude and appreciation go to my family members for their wholehearted support and encouragement, without which I would not have been able to complete this professionally enriching and personally satisfying journey.

—Debesh Mishra, PhD

Preface

This book is a series organized for easy portability and reading. It includes nine chapters. Chapter 1 discusses armor-grade composites for ballistic protection. In the research, SolidWorks 2011 is used to create a 3D fabric model, and the impact resistance of such a model is investigated using two types of bullets: first a copper bullet and subsequently a lead bullet, both at differing initial bullet speeds. In Chapter 2, the development, characterization, and application of hybrid natural fiber-reinforced polymer composites are analyzed and discussed. Chapter 3 presents a detailed study on optical fibers, their dyeing principles and methods, experimental investigation of fiber dyeing, measurement of side emission, and cross-sectional analysis of plastic optical fibers. Chapter 4 discusses how the selection of fabric characteristics significantly impacts fabric properties. As the structure is influenced by fabric properties, the behavior and interrelationship of fabric parameters (FPs) are prerequisites for finding effective solutions to fabric engineering challenges. Many textile aspects are fundamentally determined by geometrical connections. Fabric geometrical models provide simplified equations and specific constants used in textile engineering to address structural complexities and mechanical properties. These FPs are useful tools for creative fabric designers producing materials for a variety of uses. Due to the theoretical links between the FPs, the designer can experiment with different fibers, yarn tex, threads per centimeter, and weaves to modify texture and FPs.

Chapter 5 addresses quality assurance and quality control using machine learning. It discusses how machine learning is used to ensure quality assurance and reliability, helping maintain specification limits and customer satisfaction in mass production, manufacturing, marketing, transport, weather forecasting, and more. In Chapter 6, an attempt is made to assess the degrading factors responsible for reducing the mechanical performance of fiber-reinforced polymer composites. Chapter 7 discusses friction stir processing (FSP) and its ability to change the properties of metals through intense, localized plastic deformation. In this study, a surface composite based on AA7075 alloy with zinc as a primary alloying element is fabricated using FSP. The resulting composite was found to have improved hardness and corrosion resistance, along with slightly reduced strength. Chapter 8 analyzes the viscosity data for high Al_2O_3 BF slag as affected by the presence of major components CaO, SiO_2, and MgO.

CHAPTER 1

Computational Modeling of 3D Fabrics for Ballistic and Stab Protective Uses

RAJESH MISHRA*

Department of Material Science and Manufacturing Technology, Faculty of Engineering, Czech University of Life Sciences Prague, Prague-Suchdol, Prague, Czech Republic

*Corresponding author

ABSTRACT

Lightweight armored vehicles, transportable shelters, patrol boats, helicopters, and bullet-protective armor systems increasingly use armor-grade composites for ballistic protection. In this research, SolidWorks 2011 was used to create a 3D fabric model. The model is designed to be easily incorporated into commercial finite element packages such as ANSYS. The impact resistance of the model is investigated using two types of bullets—copper and lead—both at different initial velocities.

1.1 INTRODUCTION

Several lightweight ballistic protection armor systems are now comprised of composites labeled ballistic-grade or armor-grade, which are often made of high-strength polymeric fibers with excellent impact resistance, such as oriented polyethylene fibers (Spectra®, Dyneema®, and so forth), aramid (Kevlar®, Twaron®, and so forth), or glass fibers (E-Glass, and so forth). The fibers are inserted in the resin/polymer matrix, either as woven textiles or as 0?/90? cross-plied collimated continuous filaments. To make the most of the fibers' inherent high transverse impact resistance, the polymer-matrix percentage is normally kept around 20% by volume.

Advances in Industrial Manufacturing and Materials: Integrating a New Era of Computational Modeling and Machine Learning. Debesh Mishra and Swagatika Mishra (Eds.)

DOI: 10.1201/9781779640475-1

These composites are flexible to reasonably high laminate thicknesses due to the very low resin content. Hybrid constructions, in which a strong metallic or ceramic strike plate is bonded to the front of an armor-grade composite laminate, can further strengthen the armor-grade composites' penetration resistance. The current work uses finite element modeling to simulate the geometric and micromechanical properties of 3D orthogonal fabrics. For each simulation, the steps listed below are used.

1.1.1 FINITE-ELEMENT METHOD (FEM)

In practice, the finite-element method (FEM), also known as finite-element analysis (FEA), is a numerical approach for obtaining approximate solutions to partial differential equations (PDEs) and integral equations. The two possible approaches for solving the problem are to either eliminate the differential equation entirely (which addresses stationary-state problems) or to transform the PDEs into a system of ordinary differential equations, which are then numerically integrated using widely used methods such as Euler's method, Runge–Kutta, and so on. The fundamental challenge in solving PDEs is to construct an equation that closely resembles the problem under examination while remaining numerically stable, which implies that errors in the input and intermediate calculations do not accumulate and render the eventual output invalid.

1.1.2 ANSYS

An array of mechanical issues can be numerically solved using the finite element modeling tool known as ANSYS. These obstacles include static or dynamic structural analysis (both nonlinear and linear), heat transport and fluid difficulties, and electromagnetic and acoustic concerns. It uses a preprocessor software engine to produce geometry. A solution process then loads the meshed geometry, and postprocessing generates the appropriate results.

1.1.3 YARN-PATH IN 3D ORTHOGONAL FABRICS

Three perpendicularly oriented directional threads make up this fabric. In addition to layers of warp and weft yarn, there is a third layer of Z-binding yarn that joins the warp and weft layers together. A 3D orthogonal fabric's basic yarn arrangement is shown in Figure 1.1.

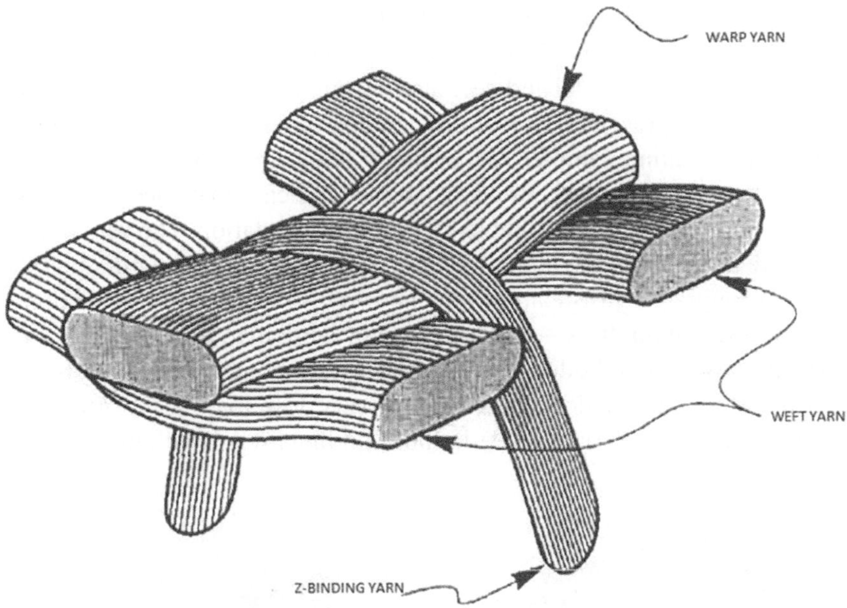

FIGURE 1.1 The yarn-path in 3D orthogonal fabrics.

The overall structure and unit cell are depicted in Figures 1.2(a) and (b). The study was carried out using a 3D orthogonal fabric unit cell. The fabric is made up of three stuffer yarn layers, four feeder yarn layers, and two binder yarn layers. The racetrack of choice is the yarn cross-section since it yields the best results in terms of application. Because the number of components in an entire fabric would be too large for ANSYS to converge on, the analysis is done on a unit cell rather than on an entire fabric. As a result, to achieve realistic conditions, a fabric's unit cell is selected for analysis with acceptable border conditions.

1.1.4 DIMENSIONS IN GEOMETRY

The unit cell's dimensions are 6.3 mm (height) × 4 mm (width) × 2.7 mm (length).

Stuffer yarns' effective number = 6
Filler yarns' effective number = 8
Binder yarns' effective number = 2

Tow fineness (linear density) = 600 Tex

Binder ends per cm = 2; Stuffer ends per cm = 6; Picks per cm = 8

To simplify the model and decrease the number of nodes and elements, the yarns are represented as isotropic monofilaments with the same linear density as genuine samples; accordingly, they are shown in the model figures as monofilaments. Previous studies indicate that racetrack cross-section modeling is more suitable for practical validation. The yarn dimensions are as follows: the major axis is 1 mm long, while the minor axis is 0.7 mm long. The justification for choosing a racetrack cross section is that in other geometries—such as circular, lenticular, and elliptical—the contact between two bodies is a line, the area of contact is extremely low, and therefore the tension caused under very low load is too great (*see* Figures 1.3(a) and (b)); as a result, those findings would be unreliable.

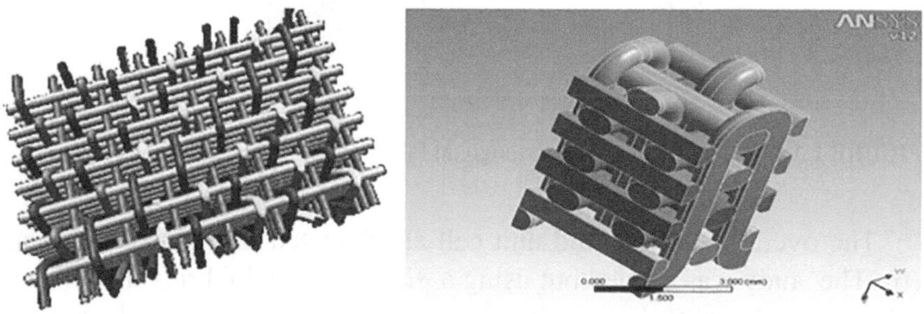

(a) (b)

FIGURE 1.2 (a) Overall structure of 3D orthogonal fabric; and (b) unit cell structure.

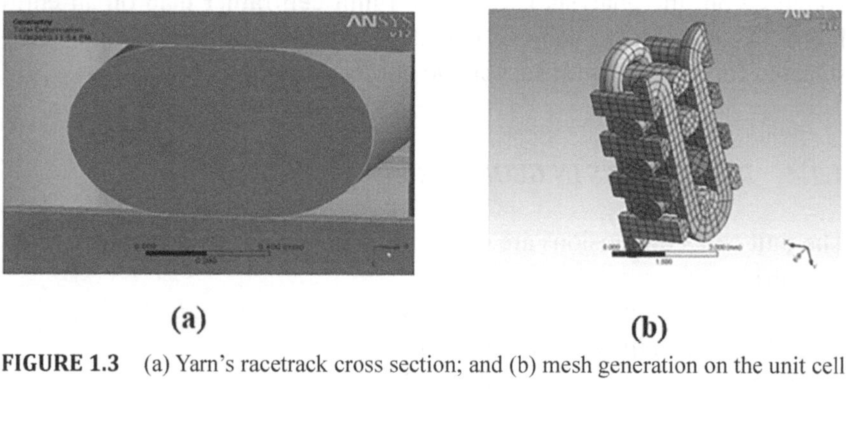

(a) **(b)**

FIGURE 1.3 (a) Yarn's racetrack cross section; and (b) mesh generation on the unit cell.

Two-dimensional (2D) and three-dimensional (3D) structures can be found in woven fabrics used in composites. 2D weaving is a reasonably quick and cost-effective procedure. Yarns are placed in a single plane in 2D construction (Figure 1.4), resulting in a single-layer design with various interlacement orders such as plain, satin, twill, and so on (Fukuta et al., 1982; Khokar, 1996). Because the fabric thickness is minimal, it is employed in laminating. However, the interwoven threads in 2D woven fabrics exhibit inherent crimp or waviness, which is undesired for optimal composite properties. From the standpoint of composite applications, the main disadvantages of 2D fabric are its anisotropy, poor in-plane shear resistance, and lower modulus than the constituent fiber material due to crimp prevalence (Khokar, 2001).

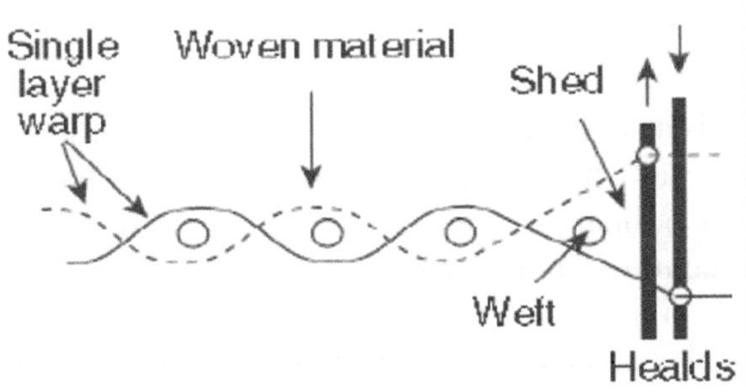

FIGURE 1.4 2D weaving processes.

Woven textiles with three-dimensional fiber architecture, on the other hand, are of special relevance for composite materials due to their high impact resistance. The structure of a 3D woven single-fabric system is highly conformable to complicated mold forms (*see* Figure 1.5). It offers consistent thickness, avoids multiple plies of 2D fabric, and is very economical in composite production. The fabric has a high fiber volume fraction and a stronger, lighter structure due to the presence of straight, noncrimp yarn (Chou and Chen, 1995; Khokar, 1997; Lee et al., 2002; Mohamed and Wetzel, 2006). Z-fibers (fibers in yarns that are oriented in the Z-direction) increase impact resistance, inhibit delaminating, wet out faster in molding, and facilitate vacuum infusion. However, many 3D weaves, such as interlock textiles, exhibit poor in-plane performance due to high warp-yarn waviness. Orthogonal weaving can be used to overcome this limitation. When warp, weft, and Z-yarns are oriented perpendicular to one another in straight paths, strength and stiffness are equivalent to those of standard plain- and twill-woven fabrics (Chen and Zanini, 1997; Khokar and Peterson, 1999).

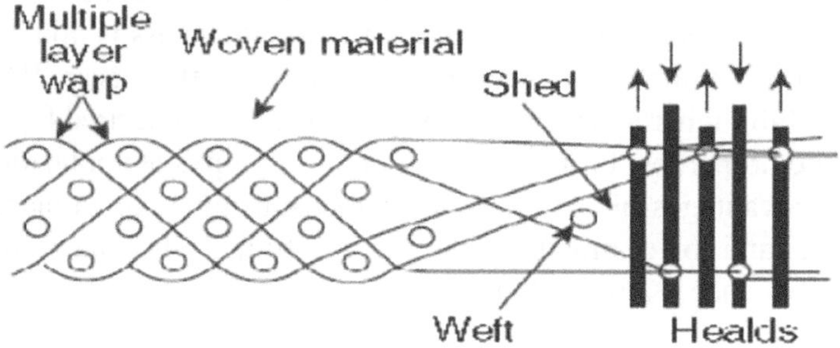

FIGURE 1.5 3D weaving processes.

Thickness is an essential criterion in 3D fabrics. Ordinary fabrics have length, width, and thickness; however, 3D fabrics have much greater thickness than ordinary fabrics. Thickness is achieved by producing a multilayer structure with multiple warp and weft layers. There are several methods for forming 3D solid architectures, each with its unique structural and mechanical characteristics. Here, three basic constructs are presented that may be produced using a 2D weaving machine with minimal modifications (Chen et al., 1999; Kuo et al., 2007).

To begin, a 3D solid architecture may be created using the multi-layer approach, in which multiple layers of various woven materials are sewn together during the weaving process (Figure 1.6). The characteristics of this type of 3D architecture are as follows:

i. Each fabric layer may be given a different weave to enhance hybrid qualities throughout the thickness of the composites;
ii. All warp and weft yarns in the 3D solid textile architectures may be crimped to specific extremes to satisfy the composite's property requirements;
iii. To improve the through-the-thickness property, the required amount of vertical stitching can be provided between any layers of fabric in the preform; and
iv. Straight wadding yarns added to any adjacent fabric layer in the warp, weft, or both directions can further improve the architecture.

FIGURE 1.6 Multilayer weaving principle.

A distinct separation of fabric layers throughout the fabric thickness distinguishes multilayer fabrics. Each layer is made up of warp ends and weft threads. To weave the layers together, existing yarns (self-stitching) or external sets of yarns are employed (central-stitching). Because of structural characteristics, all warp and weft threads in a multilayer fabric are crimped, resulting in low initial modulus under tensile loads along the warp and weft (Chen and Potiyaraj, 1999a, b; Khokar, 2002; Tung and Jayaraman, 1991).

The orthogonal principle is the second approach for creating 3D solid woven architecture, in which several warp and weft series connect at a consistent 90? angle, as in traditional textile weaving (Figure 1.7). Other features include:

i. All yarns in the three principal directions are placed straight, enabling them to bear loads directly and most efficiently;
ii. The binding-weave specification arranges the required number of vertical yarns; and
iii. Interwoven depth within the same preform can be easily varied to produce different preform sizes.

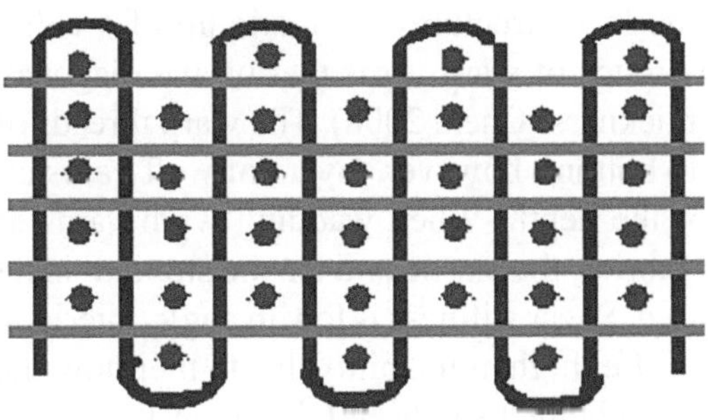

FIGURE 1.7 The 3D orthogonal fabric.

The number of warp or weft yarn layers determines the thickness of the orthogonal structures, with weft yarns having one more layer than warp yarns. In the warp, weft, and through-the-thickness orientations, yarns frequently run straight. Figure 1.8 depicts several variations of orthogonal structures.

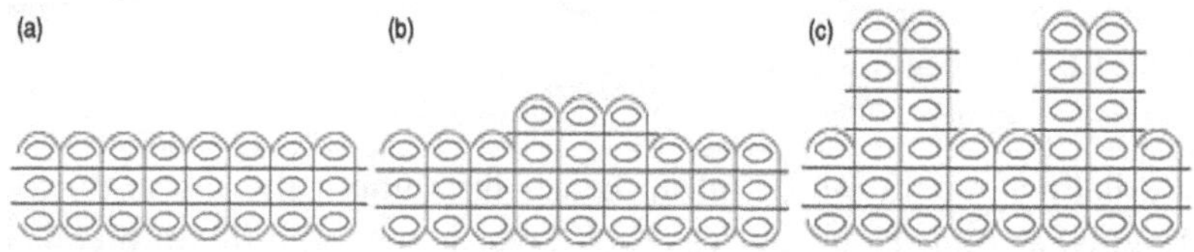

FIGURE 1.8 Variations of the orthogonal structures.

The employment of the angle-interlock concept—where weft threads are laid straight while warp yarns flow diagonally—is the third method for creating 3D architectures (Figure 1.9). The Z-yarns in this structure are not interlaced at 90? as in an orthogonal sample but are aligned at an angle to the normal. Due to fewer cross-over contact points, this topology often leads to more flexible 3D structures (Khokar, 2002; Tung and Jayaraman, 1991).

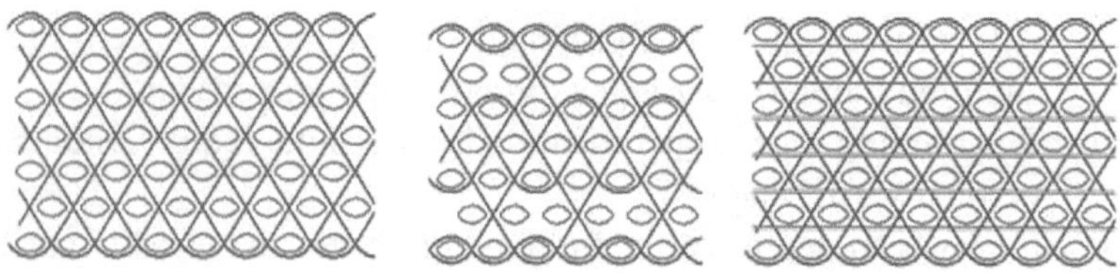

FIGURE 1.9 3D angle-interlock woven structure.

The angle-interlock structures are made up of a collection of straight weft yarns and groups of warp yarns that weave diagonally with the weft throughout the thickness (Chen, 2008). The warp threads often bond diagonally from top to bottom; however, layers of weft yarns can also be bound for a fraction of the depth. When wadding warp yarns are introduced to angle-interlock fabrics, the elastic and tensile strength in the warp direction is greatly increased. Shear stiffness is low in angle-interlock textiles. Angle-interlock textiles offer high moldability due to their low shear stiffness, and applications include helmet shells and feminine armors. Tapered preforms are produced owing to alterations in the fabric's interlocking depth (Adanur and Allen, 1995; Wolff, 2004; www.3tex.com, 2008).

A three-dimensional fabric has three mutually perpendicular planes in which the component yarns are organized. Fiber orientation in conventional 2D woven textiles is exclusively in the 0? and 90? directions. The degree of yarn crimp can be varied depending on fabric manufacture and weave pattern. Crimp levels must be kept to a minimum for efficient translation of yarn properties into fabric properties. The longer the floats in a fabric design—such as twills and satins—the lower the yarn crimp and the greater the fabric modulus and strength, particularly with filament yarn when fabric support is not critical (Antonio, 1999; Bogdanovich and Pastore, 1996; Flemming et al., 1999; Mallick, 1997; Vigo and Turbak, 1991).

Fiber-reinforced composite materials offer a number of benefits and are widely employed in a range of applications. Interlaminar delamination is the weakest failure mechanism in such materials, since the bulk of them are two-dimensional laminated composites. Several strategies for improving interlaminar strength have already been proposed, with three-dimensional textile technology being the most successful (Khokar, 1999; 2001). Numerous 3D textiles have also been produced, each with its own set of features. The orthogonal fabric composite, for example, presents a fiber architecture that aims to maintain in-plane performance while boosting interlaminar toughness by inserting a modest quantity of through-thickness (Z) reinforcements. The stages of the meso-FE analysis and the sequence for modeling such structures are:

- Creating a model of the reinforcement's internal geometry;
- Geometry modeling using a solid modeler such as ANSYS Multiphysics;
- Mesh preparation: splitting the solid into separate pieces, correcting interpenetration of yarn volumes, and making room for thin matrix layers between the yarns;
- Determining the local material properties of the impregnated yarns and matrix;
- Applying loads and producing results through deformed shapes, contour plots, nodal solutions, and so on.

Three orientations exist for the strands in this fabric, with the yarn perpendicular to each other in all three directions. The Z-binding yarns (yarns in the z axis) bind all layers of warp and weft yarn. The best feature of this type of fabric is that the yarns are not crimped, which increases the fabric's Young's modulus. Z-binding yarn has only slight crimp at the fabric edge, which can be overlooked (Budgen, 1995; Fleury et al., 1990; Lomov et al., 2008). Mechanical test data are compared to the effective elastic characteristics and strain fields predicted by two three-dimensional numerical analytical methods. The capacity of each analytical approach to predict distinct mechanical characteristics is outlined (Bogdanovich et al., 2009). For a typical structurally stitched multilayer preform made of quasi-UD carbon-fiber woven fabric, the experimental data, geometrical models, and FEM analysis are provided. The word structural implies that the consolidation of plies is not only performed by the stitching yarn as a nonstructural yarn, but also serves as a through-thickness reinforcement. One of the stitching techniques, tufting, is researched using 67 Tex carbon yarn. The models take into consideration general aspects of the local preform geometry and are thought to be adequate for modeling at the meso-scale (textile unit-cell) level. Experimental and theoretical results are given, contrasted, and analyzed; a roadmap for modeling structurally stitched preforms is offered (Perie et al., 2009). A predicted strength study of noncrimp 3D orthogonal weave composites that has been experimentally validated is provided. Progressive failure analysis using the 3D MOSAIC algorithm and tensile testing are part of the combined theoretical and experimental methodology, reinforced by acoustic emission monitoring and progressive crack detection. Theoretical and experimental results for 3D orthogonal weave composites agree quite well. WiseTex software was created to simulate 3D interlock textiles. Carbon tow compression behavior is monitored and utilized as an input. Characterization of microstructure and comparison of mechanical characteristics between trials, as well as homogenization calculations using inclusion models, validate the model (Badawi, 2008; Carvelli et al., 2009).

Due to their cost-effectiveness, high strength, environmentally sound use of materials and methods, and ability to lower product weight, the demand for novel lightweight materials has grown dramatically in recent years. Materials that can significantly improve travel speed and fuel economy while meeting consumer expectations, military objectives, and regulatory constraints are of special relevance for transportation applications. However, the demand for novel lightweight materials extends well beyond transportation. Lightweight materials and structures are becoming increasingly popular for body armor and protection, architecture and construction, mechanical engineering applications, sports and leisure items, packaging, and publication. This competition is important because it focuses on the creation of lightweight materials and structures. It can be accomplished by applying innovative design strategies to reduce the bulk of the material or the weight of the final structures. The primary goal of a Dresden research project was to create a new type of woven spacer fabric for low-weight composite materials, specifically weaving spacer textiles that cannot be accomplished using existing methods (Badawi, 2008).

Meso-scale finite-element modeling of textiles and fabric composites (fabric-structure unit-cell size) is a valuable tool for homogenizing mechanical properties, investigating stress-strain fields inside the unit cell, establishing damage-initiation conditions and locations, and simulating damage progression and deterioration of the composite's homogenized mechanical characteristics. The current study focuses on modeling, particularly on issues such as creating finite-element simulations of fabric structures (geometry, meshing, and boundary conditions), understanding the results, and validating the models. In the finite-element solution, there are three phases that form the basis for all finite-element analyses.

The analysis is performed on a unit cell rather than a whole fabric, since the number of elements in a full fabric would be prohibitively large and ANSYS would be unable to converge the solutions. As a result, a fabric unit cell is analyzed with appropriate boundary conditions to obtain realistic conditions. Figures 1.10(a) and 1.10(b) depict the overall structure and unit cell structure.

(a) (b)

FIGURE 1.10 (a) Overall structure of 3D orthogonal fabric; and (b) unit cell structure.

The structure is broken into smaller elements for analytical purposes. In Figure 1.11, mesh statistics consist of 2,273 elements and 13,318 nodes.

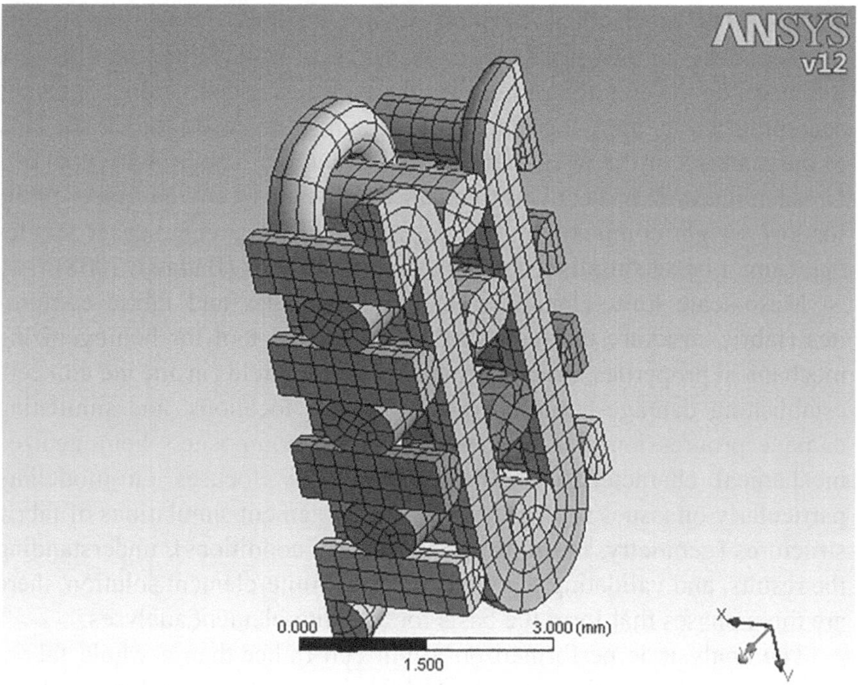

FIGURE 1.11 Mesh generation on the unit cell.

The mesh arrangement has no effect on finite element loads. The element mesh can be changed without influencing the loading, which permits mesh alterations and mesh-sensitivity tests to be performed without having to reapply loads every time. As a result, choosing a model and applying loads to the components is simpler compared to other techniques. The active element coordinate system contains elements created by ANSYS meshing procedures. Meshing instructions produce nodes that utilize the global Cartesian coordinate system. Consequently, multiple coordinate systems and loading orientations may be used with the finite element model.

When the tensile force is applied in the longitudinal direction to a 3D fabric composite system, lateral compression occurs. The compressive load is applied to stuffer yarns in this situation. Pressure load is preferable for analytical reasons to a point load among the load options available in finite element analysis, so that a uniform force may be distributed on each stuffer yarn. As indicated in Figure 1.12, a pressure load is applied to one of the faces.

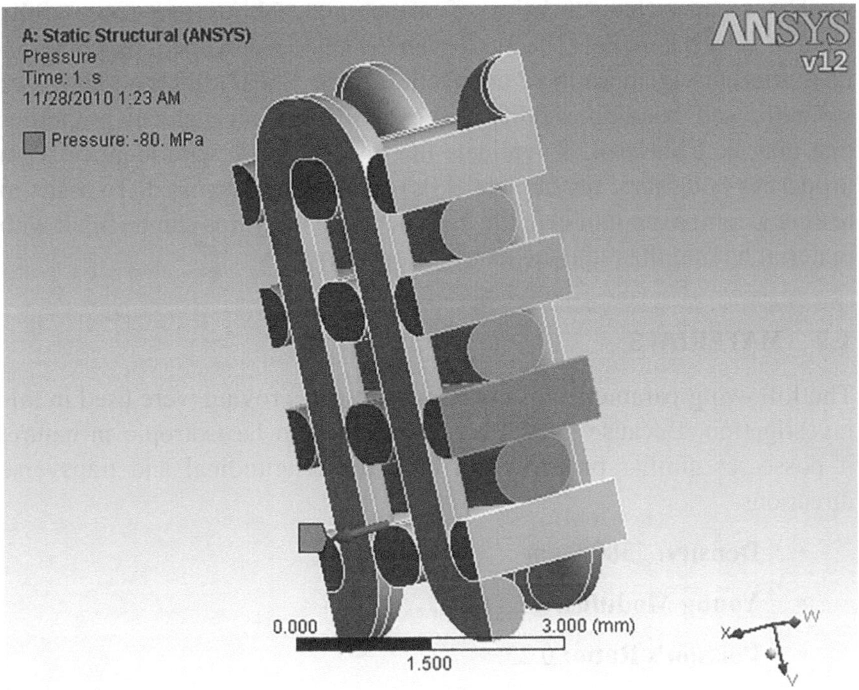

FIGURE 1.12 Pressure load applied to one of the faces.

Figures 1.13(a) and 1.13(b) illustrate the unit cell's stress distribution produced from an ANSYS simulation and the unit cell's overall deformation.

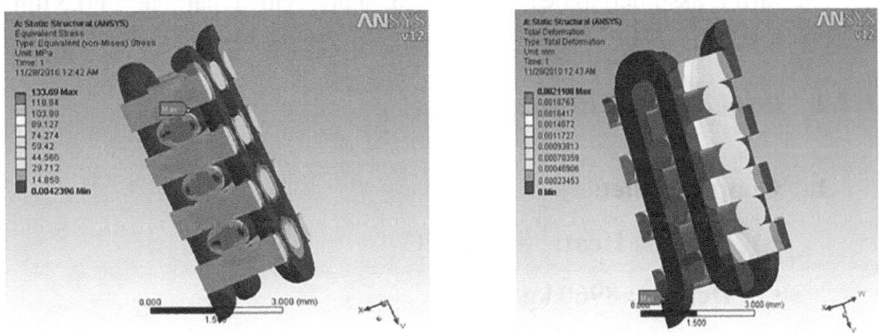

(a) (b)

FIGURE 1.13 (a) Unit cell stress distribution; and (b) unit cell overall deformation.

The tensile characteristics of orthogonal fabrics are successfully simulated by the model. The model can be enhanced by including fiber-to-fiber friction via appropriate connection parts. Shear stress, compressive strength, and bending rigidity are examples of mechanical properties that may be examined. To validate the results, fabric with identical yarn properties and fabric parameters (FPs) can be manufactured. To assist in testing a composite unit cell, the gaps between the yarns can be filled with material having the requisite matrix properties.

1.2 MATERIALS

The following parameters of multifilament glass roving were used in this investigation. Because glass fiber is assumed to be isotropic in nature, it possesses similar properties in both the longitudinal and transverse directions.

- **Density:** 2580 kg/m^3;
- **Young Modulus:** 72.3 GPa;
- **Poisson's Ratio:** 0.23;
- **Shear Modulus:** 29.39 GPa;
- **Bulk Modulus:** 44.63 GPa.

Assuming the yarn is multifilament with a packing fraction of 0.6:

$$\text{Yarn density} = \text{Fiber density} \times \text{Packing fraction} = 2{,}580 \times 0.6 = 1{,}548 \text{ kg/m}^3$$

The yarn is assumed to be of cylindrical shape with a diameter of 0.5 mm.

1.2.1 BULLET MATERIAL

1. **Copper Bullet:**
   - **Specific Heat:** 383 J/kg/°C
   - **Density:** 8960 kg/m^3
   - **Bulk Modulus:** 12.9×10^{10} Pa
   - **Shear Modulus:** 46×10^9 Pa

2. **Lead Bullet:**
    - **Specific Heat:** 124 J/kg/°C
    - **Density:** 11,340 kg/m^3
    - **Shear Modulus:** 8.6×10^9 Pa

1.2.2 GEOMETRY OF BULLET

As illustrated in Figure 1.14, the bullet's bottom section is cylindrical, with a diameter of 1 mm, while the upper section is an elliptical dome.

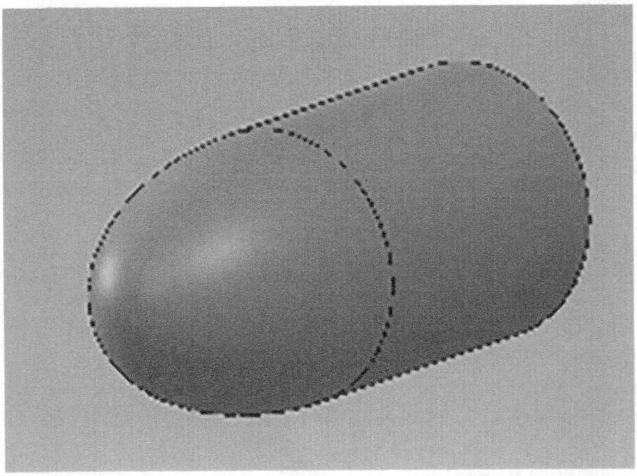

FIGURE 1.14 Bullet geometry.

1.2.3 PENETRATION TEST OF KNIFE

The Testometric Tester M350-10CT was used to perform this test. This machine has an adjustable penetration speed and interchangeable knife types. A single sharp-pointed knife, shown in Figure 1.15, was used for testing. Each fabric sample was placed on the machine's base plate, and the machine head was lowered onto the sample to hold it in place. At a speed of 1,000 mm/min, the knife pierced the sample. The equipment includes a recording monitor that captures the maximum force applied during penetration. The same test was performed on composites reinforced with various types of fly ash nanoparticles and woven glass fabrics.

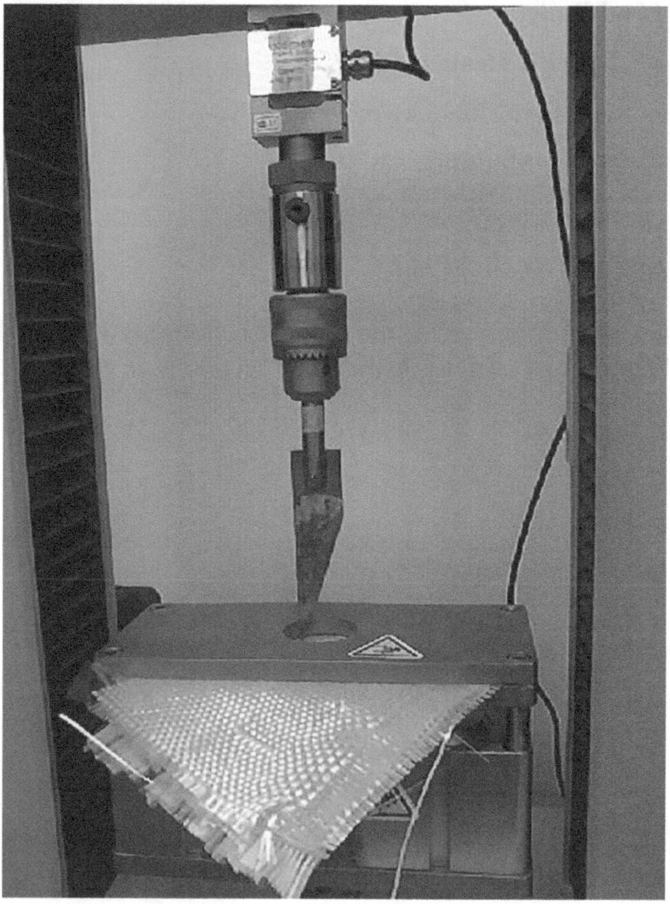

FIGURE 1.15 Testometric M350-10CT knife penetration tester.

The simulation begins with copper bullet material and varying initial bullet velocities.

- **Case 1:**
  - **Bullet Material:** Copper;
  - **Initial Bullet Velocity:** 300 m/s;
  - No fixed support (fabric sample in free suspension).

When the fabric has no support, it is entirely destroyed; however, in reality, the fabric's area is supported by surrounding yarns, so all sides of the fabric except the one exposed to the bullet have fixed support (Figure 1.16).

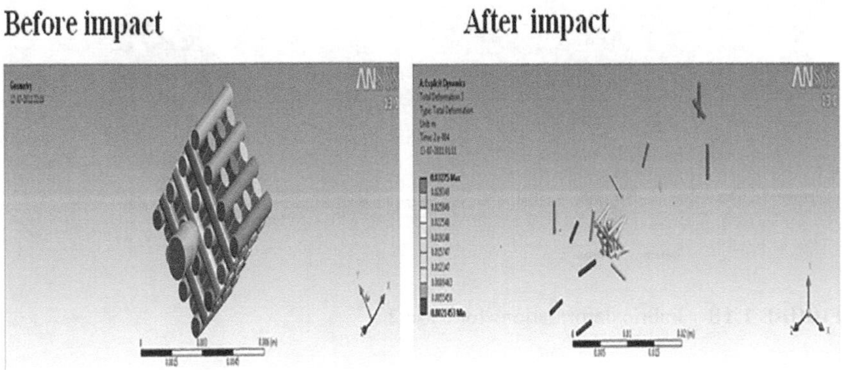

FIGURE 1.16 Fabric deformations for Case 1.

➢ **Case 2:**

- **Bullet Material:** Copper;
- **Initial Bullet Velocity:** 300 m/s.
- **Loading and Boundary Conditions:** Except for the face exposed to the bullet, all faces have fixed support (*see* Figures 1.17–1.19).

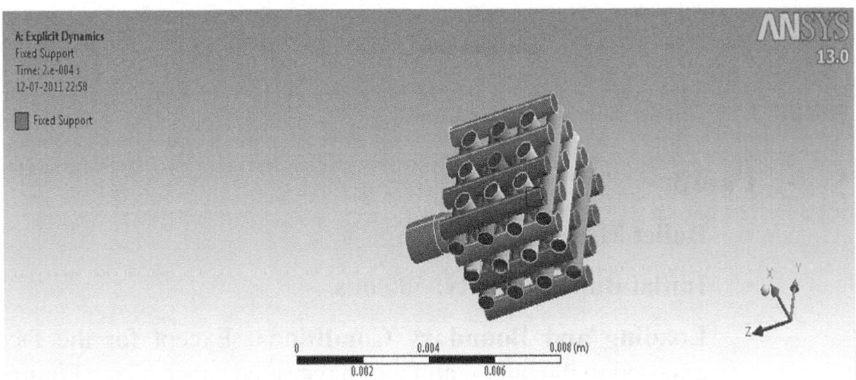

FIGURE 1.17 Fixed support for the unit cell.

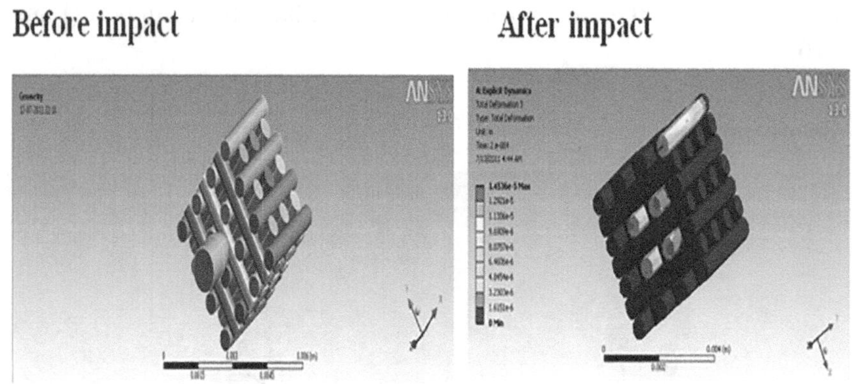

FIGURE 1.18 Fabric deformations for Case 2.

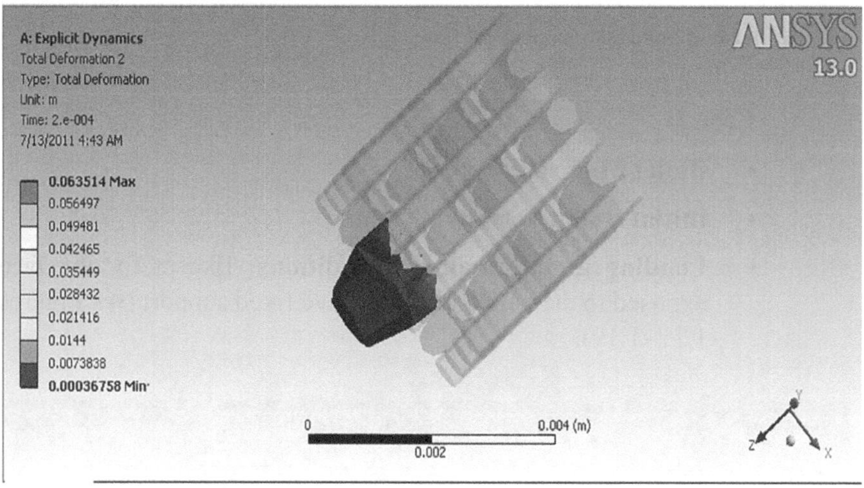

FIGURE 1.19 Bullet deformations for Case 2.

- **Case 3:**
  - **Bullet Material:** Copper
  - **Initial Bullet Velocity:** 500 m/s
  - **Loading and Boundary Conditions:** Except for the face exposed to the bullet, all faces have fixed support (*see* Figures 1.20 and 1.21).

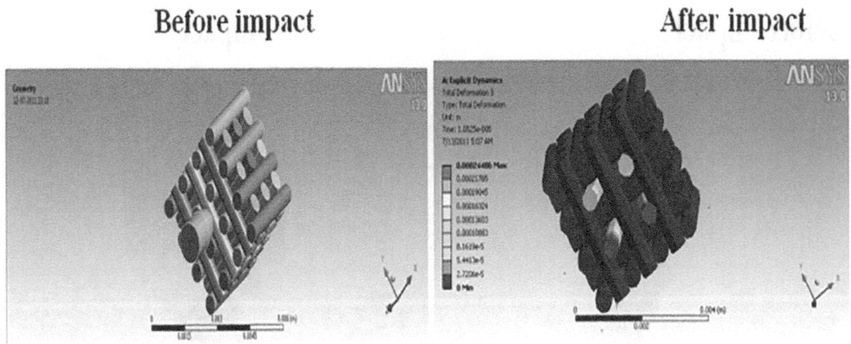

FIGURE 1.20 Fabric deformations for Case 3.

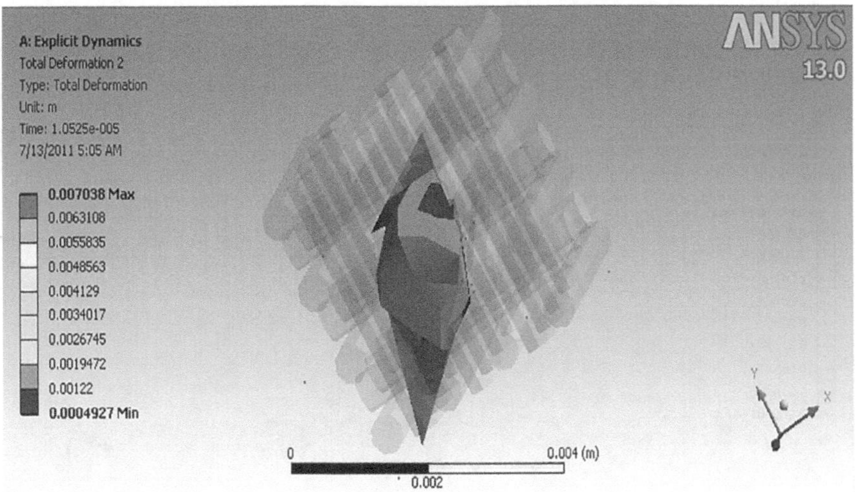

FIGURE 1.21 Bullet deformations for Case 3.

➢ **Case 4:**

- **Bullet Material:** Copper.
- **Initial Bullet Velocity:** 700 m/s.
- **Loading and Boundary Conditions:** Except for the face exposed to the bullet, all faces have fixed support (*see* Figures 1.22 and 1.23).

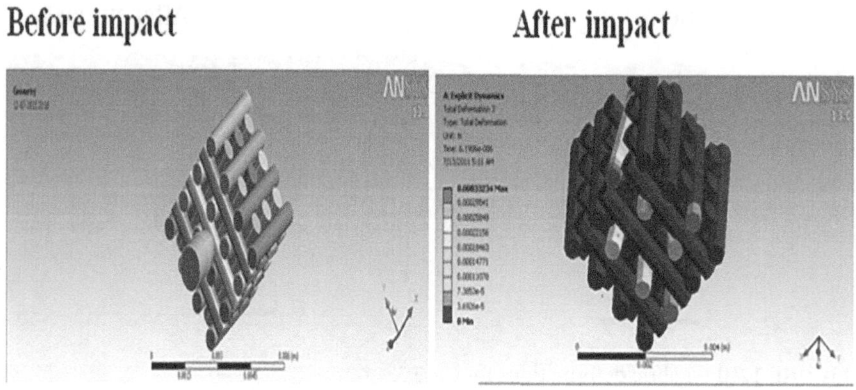

FIGURE 1.22 Fabric deformations for Case 4.

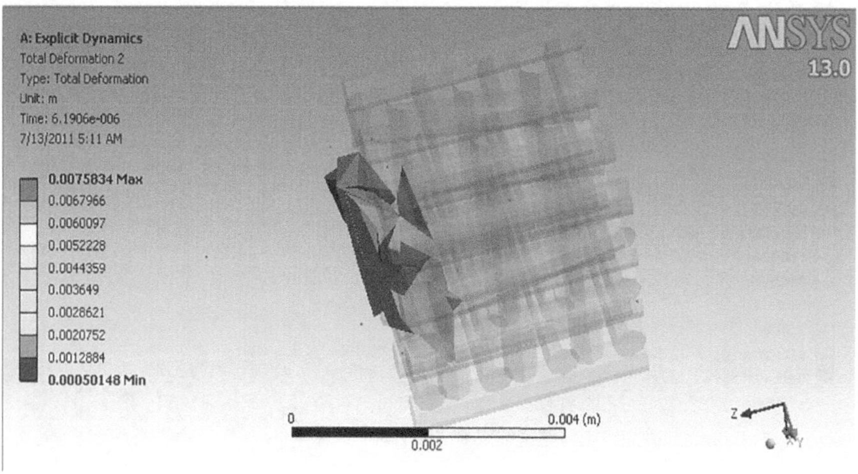

FIGURE 1.23 Bullet deformations for Case 4.

- **Case 5:**
  - **Bullet Material:** Copper
  - **Initial Bullet Velocity:** 900 m/s
  - **Loading and Boundary Conditions:** Except for the face exposed to the bullet, all faces have fixed support (*see* Figures 1.24 and 1.25).

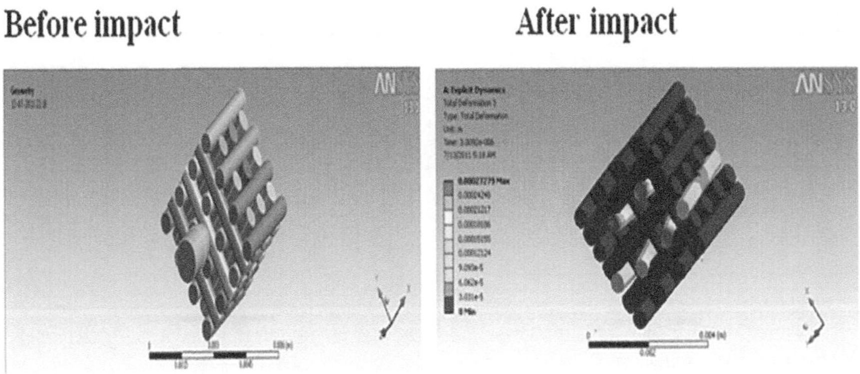

FIGURE 1.24 Fabric deformations for Case 5.

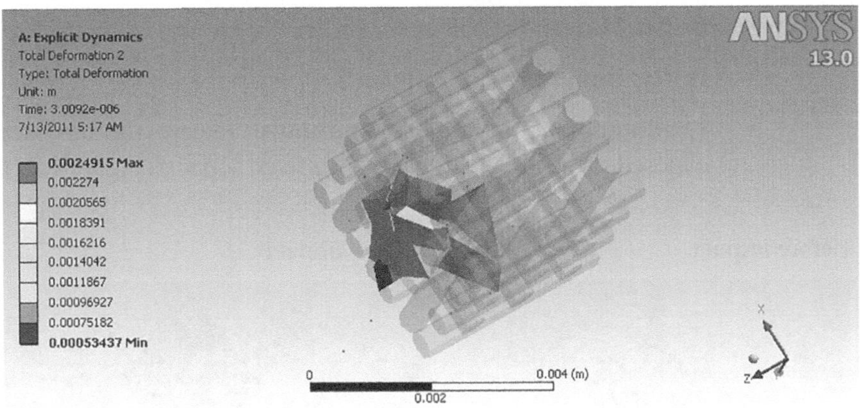

FIGURE 1.25 Bullet deformations for Case 5.

For the impact simulation, the second type of bullet with a lead composition was used. Lead is significant in its own right because it is exceedingly dense, producing a high mass and consequently high kinetic energy for a given volume. Lead is also inexpensive, readily available, and easy to work with. It melts at relatively low temperatures, allowing for simple bullet production.

- **Bullet Material:** Lead
- **Initial Bullet Velocity:** 300 m/s
- No fixed support (fabric sample in free suspension) (Figure 1.26).

Before impact **After impact**

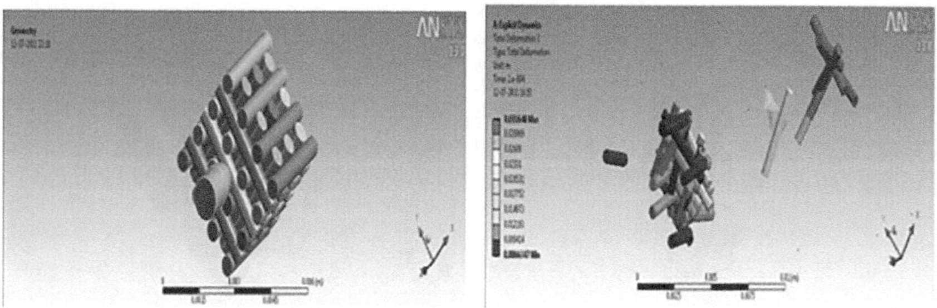

FIGURE 1.26 Fabric deformations for Case 6.

- **Case 7:**
  - **Bullet Material:** Lead
  - **Initial Bullet Velocity:** 300 m/s
  - **Loading and Boundary Conditions:** Except for the face exposed to the bullet, all faces have fixed support (Figure 1.27).

Before impact **After impact**

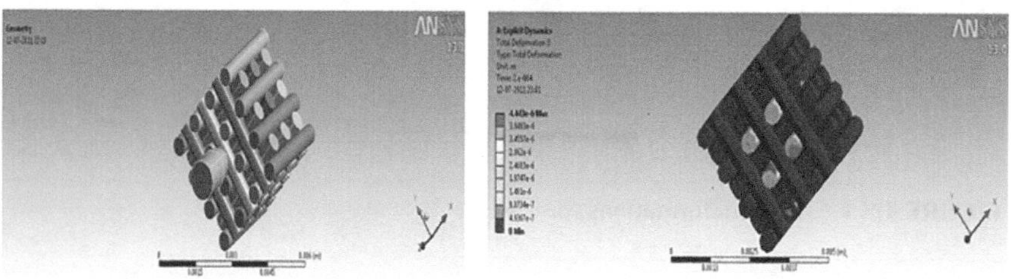

FIGURE 1.27 Fabric deformations for Case 7.

- **Case 8:**
  - **Bullet Material:** Lead
  - **Initial Bullet Velocity:** 200 m/s
  - **Loading and Boundary Conditions:** Except for the face exposed to the bullet, all faces have fixed support (Figure 1.28).

Before impact **After impact**

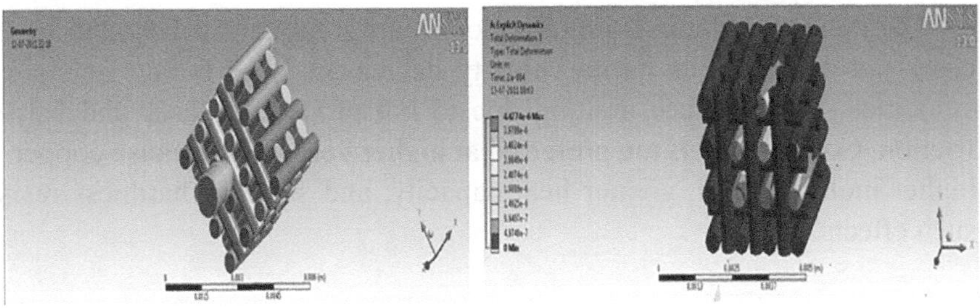

FIGURE 1.28 Fabric deformations for Case 8.

- **Case 9:**
  - **Bullet Material:** Lead
  - **Initial Bullet Velocity:** 100 m/s
  - **Loading and Boundary Conditions:** Except for the face exposed to the bullet, all faces have fixed support (Figure 1.29).

Before impact **After impact**

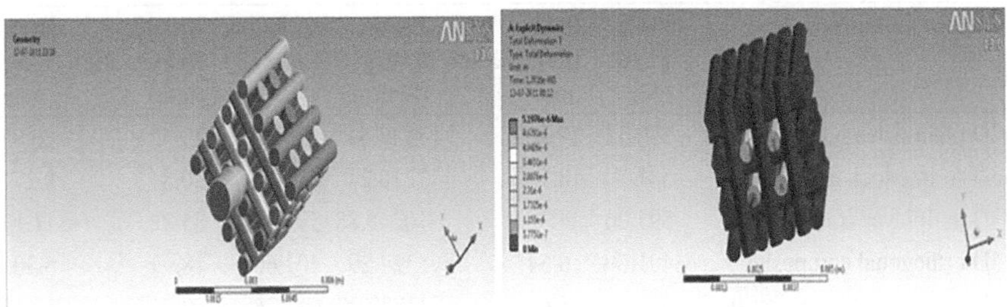

FIGURE 1.29 Fabric deformations for Case 9.

The highest bullet velocity causes the greatest fabric deformation, which is understandable given that the bullet has the most kinetic energy in this situation. The distortion of copper bullets is evident and increases as the initial velocity rises. The bullet undergoes clear deformation upon impact, with deformation magnitude growing alongside starting velocity. In the case of lead, however, the bullet is completely destroyed, as is the surrounding fabric. In modern bullets, a lead core is jacketed or plated with gliding metal, cupronickel, or copper alloys; a thin coating of harder metal protects the soft lead core during barrel travel and flight, ensuring intact delivery to the target.

The major finding is that bullet deformation remains substantial in all cases but decreases as initial velocity decreases. Lead bullets at higher velocities suffer surface melting due to hot propellant gases and barrel friction. Copper bullets are preferred at higher velocities because copper's higher melting point, greater heat capacity, and superior hardness resist such effects.

1.2.4 STAB RESISTANCE/PENETRATION TEST OF KNIFE

Table 1.1 shows the breakdown strength and work at breaking load. The breaking load of a 3D woven fabric-reinforced composite is greater than that of a 2D plain-woven composite. The 3D orthogonal fabric has the lowest resistance, which may be due to the structure's fewer interlacement sites, allowing a crack to spread readily across parallel threads when a knife penetrates the composite. The simple woven composite deforms more upon knife penetration, and the same is true for multilayer composites.

TABLE 1.1 Outcomes of Knife Penetration Test

Type of Composite	Peak Load (N)	Peak Deformation (mm)	Peak Energy (N·mm)	Breaking Load (N)	Break Deformation (mm)	Break Energy (N·m)
2D plain woven composite	530.33	19.36	7137.64	530.33	29.32	10.83
3D A. interlock composite	581.70	18.79	7210.24	581.70	20.84	8.27
3D multilayer composite	593.06	24.35	10245.88	593.06	25.82	11.17
3D orthogonal composite	401.60	16.54	5324.29	401.60	18.93	6.30
2D plain woven nanocomposite	647.26	25.69	11889.22	647.26	39.71	18.63
3D A. interlock nanocomposite	551.90	26.08	10257.22	551.90	37.86	15.07
3D multilayer nanocomposite	624.50	31.26	13338.35	624.50	51.29	21.94
3D orthogonal nanocomposite	483.93	22.96	7849.11	483.93	32.37	10.93

Figures 1.30 and 1.31 show the effect of fly ash nanoparticles. The inclusion of nanoparticles increased breakdown strength and energy by 30–40%. The breaking strength of the composites changed the most when the nanoparticles were evenly distributed on the surface and between the plain-woven fabric layers.

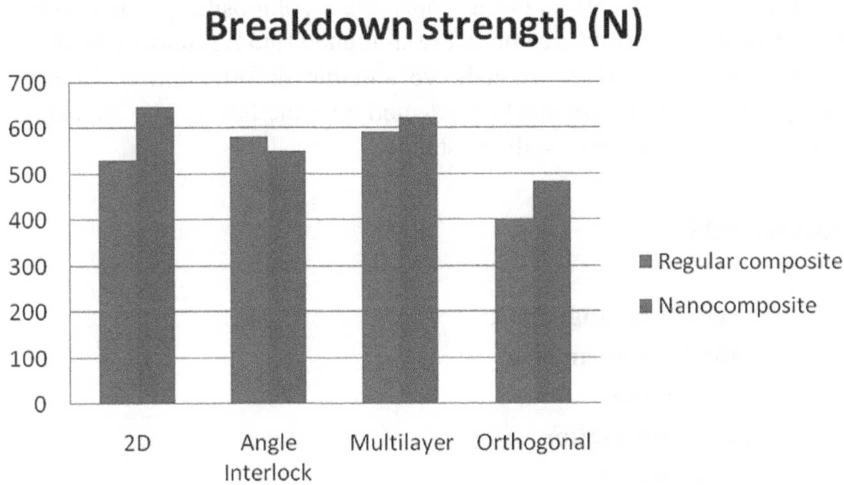

FIGURE 1.30 Effect of the addition of nanoparticles on the breakdown strength of composites.

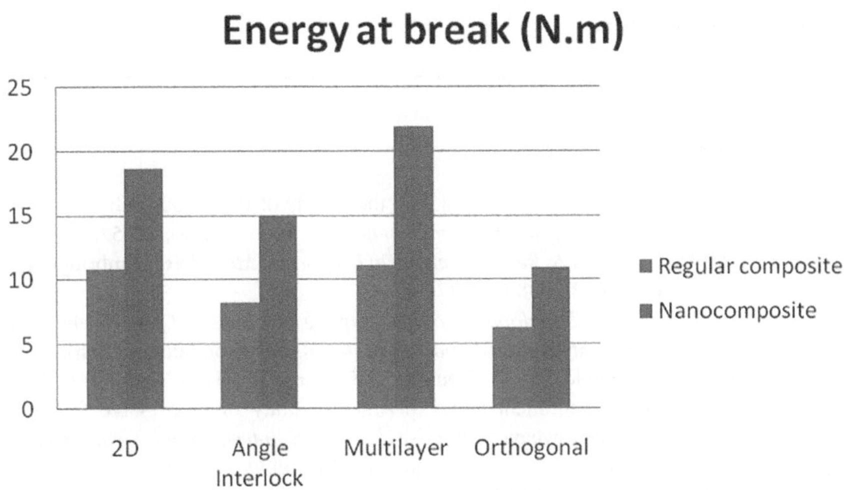

FIGURE 1.31 Effect of the addition of nanoparticles on the energy at break of composites.

1.3 CONCLUSION

Lead and copper bullets can penetrate the 3D fabric arrangement. However, the mass and velocity range is limited to simple rifle rounds. Improved fabric composition and/or structure may be necessary for bullets with higher impact. By building appropriate composite systems with the specified fabric as reinforcement, the simulated findings may be confirmed with experimental data. To achieve suitable deformation resistance, a compatible resin matrix must be selected with the fabric. The inclusion of microparticles improves stab resistance.

KEYWORDS

- **3D fabric arrangement**
- **ballistic performance**
- **copper-bullet**
- **fabric deformation**
- **kinetic energy**
- **lead-bullet**
- **nanocomposites**
- **nanoparticles**
- **stab-resistance**

REFERENCES

Adanur, S., & Allen, B. (1995). First results on the effects of ISO 9000 in the US textile industry. *Benchmarking for Quality Management & Technology, 2*(3), 41–52.

Antonio, M. (1999). *3-D Textile Reinforcements in Composite Materials*. Cambridge, UK: Woodhead Publishing. ISBN 1-85573-376-5.

Badawi, M. S. S. S. (2008). *Development of the Weaving Machine and 3D Woven Spacer Fabric Structures for Lightweight Composites Materials.* (Doctoral dissertation).

Bogdanovich, A. E., Mungalov, D. D., Lomov, S. V., Ivanov, D. S., & Verpoest, I. (2009, January). A combined theoretical and experimental study of progressive failure of non-crimp 3D orthogonal weave composite. In *17th International Conference on Composite Materials (ICCM-17), Edinburgh, IOM Communications Ltd.*

Bogdanovich, A., & Pastore, C. M. (1996). *Mechanics of Textile and Laminated Composites: With Applications to Structural Analysis*. Springer Science & Business Media.

Budgen, D. (1995). 'Design models' from software design methods. *Design Studies, 16*(3), 293–325.

Carvelli, V., Gramellini, G., Lomov, S. V., Bogdanovich, A. E., Mungalov, D. D., & Verpoest, I. (2009, January). Non-crimp 3D weave vs. laminated plane weave composites: tension-tension fatigue behavior. In *17th International Conference on Composite Materials (ICCM-17).*

Chen, X. (2008, April). CAD/CAM for 3D Fabrics for Conventional Looms. In *1st World Conference on 3D Fabrics and Their Applications*.

Chen, X., & Potiyaraj, P. (1999a). CAD/CAM of orthogonal and angle-interlock woven structures for industrial applications. *Textile Research Journal*, *69*(9), 648–655.

Chen, X., & Potiyaraj, P. (1999b). CAD/CAM for complex woven fabrics. Part II: multi-layer fabrics. *Journal of the Textile Institute*, *90*(1), 73–90.

Chen, X., & Zanini, I. (1997). An experimental investigation into the structure and mechanical properties of 3D woven orthogonal structures. *The Journal of The Textile Institute*, *88*(4), 449–464.

Chen, X., Spola, M., Paya, J. G., & Sellabona, P. M. (1999). Experimental studies on the structure and mechanical properties of multi-layer and angle-interlock woven structures. *Journal of the Textile Institute*, *90*(1), 91–99.

Chou, S., & Chen, H. E. (1995). The weaving methods of three-dimensional fabrics of advanced composite materials. *Composite structures*, *33*(3), 159–172.

Flemming, M., Ziegmann, G., & Roth, S. (1999). *Faserverbundbauweisen–Verfitungsverfahren mit duroplastischer Matrix.*

Fleury, G. A., Lavallee, R. L., & Ohnstad, T. S. (1990). *U. S. Patent No. 4,949,761.* Washington, DC: U. S. Patent and Trademark Office.

Fukuta, K., Onooka, R., Aoki, E., & Tsumuraya, S. (1982). *U. S. Patent No. 4,336,296.* Washington, DC: U. S. Patent and Trademark Office.

Gries, T. (2008). 2 Textile Reinforcement Structures Thomas Gries, Jan Stüve, and Tim Grundmann. *Ceramic Matrix Composites: Fiber Reinforced Ceramics and their Applications*. www.3tex.com (January 2008).

Khokar, N. (1996). 3D fabric-forming processes: distinguishing between 2D-weaving, 3D-weaving and an unspecified non-interlacing process. *Journal of the Textile Institute*, *87*(1), 97–106.

Khokar, N. (1997). *3D-Weaving and Noobing: Characterization of Interlaced and Non-Interlaced 3D Fabric Forming Principles*. Chalmers University of Technology.

Khokar, N. (1999). A classification of shedding methods. *Journal of the Textile Institute*, *90*(4), 570–579.

Khokar, N. (2001). 3D-weaving: theory and practice. *Journal of the Textile Institute*, *92*(2), 193–207.

Khokar, N. (2002). Noobing: A nonwoven 3D fabric-forming process explained. *Journal of the Textile Institute*, *93*(1), 52–74.

Khokar, N., & Peterson, E. (1999). An experimental uniaxial noobing device: Construction, method of operation, and related aspects. *Journal of the Textile Institute*, *90*(2), 225–242.

Kuo, W. S., Ko, T. H., & Chen, C. P. (2007). Effect of weaving processes on compressive behavior of 3D woven composites. *Composites Part A: Applied Science and Manufacturing*, *38*(2), 555–565.

Lee, L., Rudov-Clark, S., Mouritz, A. P., Bannister, M. K., & Herszberg, I. (2002). Effect of weaving damage on the tensile properties of three-dimensional woven composites. *Composite structures*, *57*(1–4), 405–413.

Lomov, S. V., Ivanov, D. S., Verpoest, I., Bogdanovich, A. E., Mungalov, D., Zako, M., Kurashiki, T,, & Nakai, H. (2008, June). Predictive analyses and experimental validations of effective elastic properties of 2D and 3D woven composites. In *13th European conference on composite materials (ECCM-13)* (p. 10).

Mallick, P. K. (1997). Composites Engineering Handbook. New York, USA: Marcel Dekker. ISBN 0-824-79304-8

Mohamed, M. H., & Wetzel, K. K. (2006*). 3D Woven Carbon/Glass Hybrid Spar Cap for Wind Turbine Rotor Blade.*

Perie, G., Lomov, S. V., Verpoest, I., & Marsal, D. (2009). Meso-scale modelling and homogenization of interlock reinforced composite. In *17th International Conference on Composite Materials (ICCM-17).*

Tung, P. S., & Jayaraman, S. (1991). *Three-Dimensional Multilayer Woven Preforms for Composites,* 53–80.

Vigo, T., & Turbak, A. (1991). High-tech fibrous materials. Washington, DC, USA: American Chemical Society. ISBN 0-8412-1985-0.

Wolff, E. G. (2004). *Introduction to the Dimensional Stability of Composite Materials.* DEStech Publications, Inc.

CHAPTER 2

In the New Era of Advanced Materials: A Critical Study

JITESH KUMAR SINGH,[1] ARUN KUMAR ROUT,[2,*] and BIBHU PRASAD DASH[3]

[1]*Department of Production Engineering, Veer Surendra Sai University of Technology, Burla, Odisha, India*

[2]*Department of Mechanical Engineering, OP Jindal University, Raigarh, Chhattisgarh, India*

[3]*Department of Textile Engineering, Odisha University of Technology and Research, Bhubaneswar, Odisha, India*

*Corresponding author

ABSTRACT

Fiber-reinforced polymer composites are increasingly used to replace metallic materials or composites, either partially or fully, in some applications. Synthetic fiber-reinforced polymer composites generally exhibit superior mechanical properties compared to natural fiber-reinforced polymer composites. However, certain natural fiber composites possess higher specific strength than their synthetic counterparts. Therefore, it is necessary to hybridize natural–natural or natural–synthetic fiber-reinforced composites (SFRCs) to compensate for the weaknesses of individual fiber types. In this chapter, the development, characterization, and application of hybrid natural fiber-reinforced polymer composites are analyzed and discussed.

2.1 INTRODUCTION

A major revolution is underway in infrastructure, transportation, marine industries, aerospace, and space manufacturing as steel is replaced

Advances in Industrial Manufacturing and Materials: Integrating a New Era of Computational Modeling and Machine Learning. Debesh Mishra and Swagatika Mishra (Eds.)

DOI: 10.1201/9781779640475-2

by composite materials. Composites are inhomogeneous, anisotropic materials formed by combining two or more constituents with different properties. Many composites exhibit superior strength and performance compared to metallic materials. Composite materials may be classified based on the geometry of their reinforcement component, which dictates their mechanical properties and overall performance (Assarar et al., 2015; Pickering et al., 2016).

Natural fibers are inexpensive and biodegradable, and their properties can be enhanced by hybridization with synthetic fibers. Manmade glass fibers can be combined with natural fibers to produce natural–synthetic hybrid composites with improved mechanical properties, high strength, reduced cost, and favorable chemical resistance. Hybrid fiber composites are used in vehicles, aircraft, civil infrastructure, marine applications, smart structures, telecommunication components, and sports goods (Kumar et al., 2015; Sanjay et al., 2018).

The two broad classifications of composites are particle-reinforced composites, or particulate composites, and fiber-reinforced composites, or fibrous composites. Fiber-reinforced composites are commonly grouped based on the nature of their fibers or reinforcement and the matrix. The matrix binds the fibers together, maintaining their alignment in the primary stress direction. When loads are applied to the composites, they are transferred onto the fibers. Fiber-reinforced composites are categorized into two groups: "natural fiber-reinforced composites (NFRCs)" and "synthetic fiber-reinforced composites (SFRC)." Natural fiber composites have good specific properties and are a viable alternative to glass fiber (Evans, 1992; Shalwan and Yousif, 2013).

2.1.1 NATURAL FIBER-REINFORCED COMPOSITES (NFRCS)

Natural fiber-reinforced polymer composites are eco-friendly materials that are lightweight, inexpensive, and biodegradable. Natural fibers are non-synthetic and can be derived from plants or animals. They are a cost-effective alternative to synthetic materials and offer high specific properties at a lower cost and with easy availability. Natural fibers are used in building materials, animal feed, medicine, and clothing. Although eco-friendly, natural fibers have some disadvantages, such as low impact strength, limited durability, poor fire resistance, and weak fiber–matrix bonding. By hybridizing with synthetic fibers, the properties of natural fibers can be improved (De Rosa et al., 2011; Ho et al., 2012; Jawaid and Khalil, 2011). Figure 2.1 shows the classification of fibers.

2.1.2 SYNTHETIC FIBER-REINFORCED COMPOSITES (SFRCS)

Synthetic fibers are produced through chemical processes and are light-weight and inexpensive. They are used in applications where a high strength-to-weight ratio is essential. Synthetic fibers are employed in the manufacture of carbon fiber-reinforced composites, basalt fiber-reinforced composites, glass fiber-reinforced composites, and aramid fiber-reinforced composites. These fibers are used in automotive parts, aircraft components, ship superstructures, and sporting goods. Synthetic fibers account for more than 60% of global fiber consumption (Jawaid and Khalil, 2011; Turra and Baruque-Ramos, 2017).

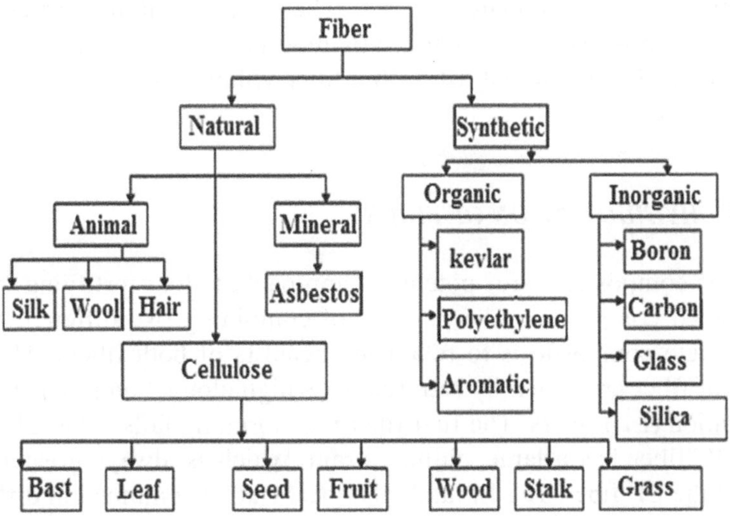

FIGURE 2.1 Classification of fibers.

Source: Reprinted with permission from Jawaid and Khalil, 2011. Copyright 2011. Elsevier.

2.1.3 HYBRID FIBER-REINFORCED COMPOSITES

Hybrid fiber composites are fabricated by combining two or more natural or synthetic fibers with a polymer matrix. Hybrid composites are more advanced than traditional fiber-reinforced polymer composites and have broad potential applications in the engineering field. Generally, they combine a high-modulus fiber with a low-modulus fiber. The high-modulus fiber contributes stiffness and load-bearing properties, while the low-modulus fiber increases damage tolerance and helps reduce material cost (Cavalcanti et al., 2019).

To form a hybrid composite, more than one discontinuous phase is embedded in a continuous phase. The discontinuous phases, known as hybrid reinforcements, are stronger and harder than the continuous phase, which is referred to as the matrix. Hybrid fibers in the composite can withstand higher loads than single-fiber reinforcement, especially in multiple directions. The matrix retains the reinforcement in the desired position and facilitates load transfer between fibers and reinforcement.

Typically, a high-modulus fiber is combined with a low-modulus fiber. The high-modulus fiber provides stiffness and load-bearing capabilities, while the low-modulus fiber improves compactness and density and reduces material cost. The mechanical properties of a hybrid composite can be enhanced by modifying the stacking sequence of different layers and adjusting the fiber volume ratio (Sathishkumar et al., 2014).

2.1.4 HYBRIDIZATION OF FIBERS

Hybrid composites are potential approaches for fortifying fiber-reinforced composites. The purpose of combining two different fibers in a single composite is to retain the benefits of both fibers. The two types of fiber are generally referred to as high elongation (HE) and low elongation (LE) fibers. The first fiber that normally fails is the LE fiber. The HE fiber has a larger failure strain, which is always greater than that of the LE fiber. The LE and HE fibers can be combined in different configurations. The three important configurations are: interlayer or layer-by-layer configuration, in which layers of two fibers are stacked onto each other; yarn-by-yarn, in which fibers are mixed together; and fiber-by-fiber, as illustrated in Figure 2.3. It is the cheapest and simplest method for producing hybrid composites (Bandaru et al., 2016; Swolfs et al., 2014). Figure 2.2 shows the effect of fiber hybrid composites: (a) apparent failure strain enhancement and (b) deviation from the rule of mixtures.

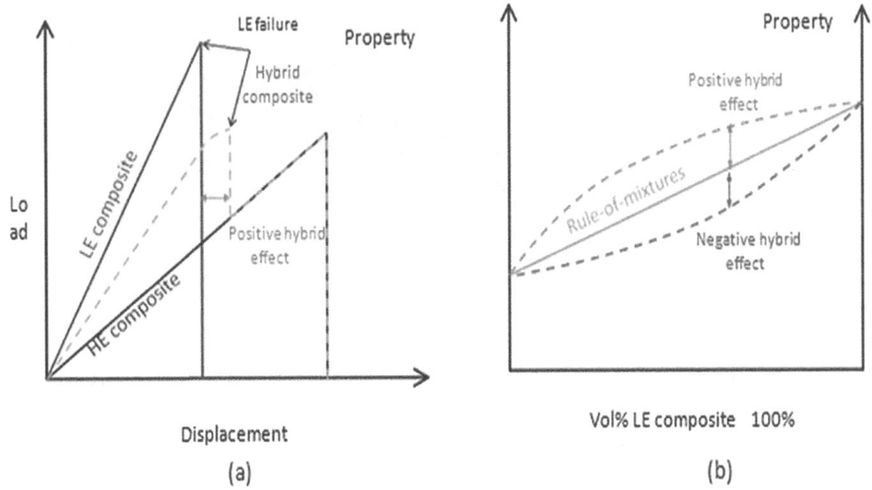

FIGURE 2.2 Effect of hybrid composites: (a) Apparent failure strain enhancement; (b) deviation from the rule of mixtures.

Source: Reprinted with permission from Swolfs et al., 2014

The intrayarn hybrid can be made by mixing at the fiber level. A more complex configuration can be obtained by combining two of these three configurations. The intrayarn layer is made by weaving the fibers together with a similar yarn. A critical aspect of hybrid composites is the dispersion of the two fibers, and the best dispersion can be achieved if the two fibers are entirely distributed. The properties of hybrid composites depend on the length of specific fibers, fiber content, matrix bonding, and arrangement of two or more fibers. The hybrid composite strength depends on the individual fibers' strain compatibility and is predicted by the mixing rule expressed in Eqn. 2.1.

$$PH = P_1V_1 + P_2V_2 \tag{2.1}$$

where; *PH* shows the property to be invented; P_1, P_2 are the corresponding properties of the first and second system, respectively; and V_1, V_2 are the hybrid volume fractions of the first and second systems, respectively (Czél and Wisnom, 2013; Swolfs et al., 2014). Figure 2.3 shows the systematic arrangement of fibers in different fashions.

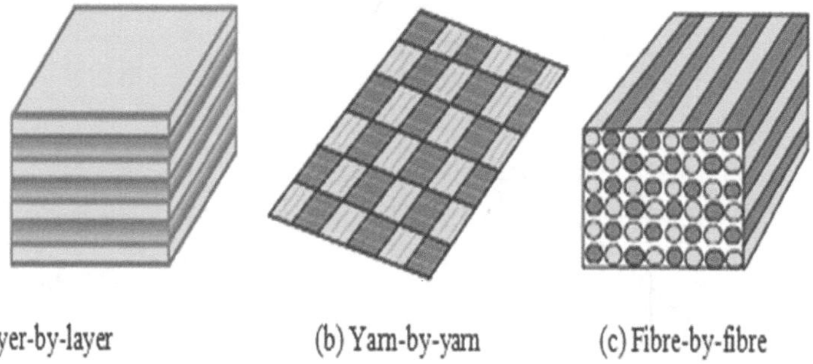

FIGURE 2.3 Hybrid configurations.

Source: Reprinted with permission from Swolfs et al., 2014. Copyright 2014. Elsevier.

2.2 METHODS OF HYBRIDIZATION OF FIBERS

The hybrid composites are selected based on the requirements and materials being designed. The hybrid composites are characterized as:

1. **Interplay or Two by Two:** Two or more constituents are mixed in a random or regular manner.
2. **Sandwich:** One material is sandwiched between the two plies or layers.
3. **Interlayer or Laminated:** Two or more materials are attached in a consistent manner.
4. **Intimately Mixed Hybrid:** In this hybrid, the constituent fibers are mixed randomly so that no overconcentration of any one type is present in the materials.
5. **Other Kind:** Such as reinforced fiber with pultruded wires, ribs, or a combination of the above.

The hybrid system for improved structure and its performance is inspired from nature, which is the recently innovative materials that are used in structural design (Ravishankar et al., 2019). A study of natural material shows how high structural performance can be achieved through hybrid combination. Hybrid composites are designed by the combination of natural fiber and synthetic fiber, or the amalgamation of two natural fibers in a matrix. The mechanical properties of natural fiber can be enhanced by hybridization with epoxy (Sarasini et al., 2017; Xie et al., 2018).

2.3 CLASSIFICATION OF HYBRID FIBER COMPOSITES

There are various types of hybrid composites, such as hybrids between natural–natural fibers, synthetic–synthetic fibers, and natural–synthetic fibers, which are used for potential applications in the engineering field. Hybrid fiber composites are broadly categorized into three groups (Amir et al., 2019; Begum and Islam, 2013). These are:

- Natural–natural fiber–reinforced hybrid composites;
- Synthetic–synthetic fiber–reinforced hybrid composites;
- Natural–synthetic fiber–reinforced hybrid composites.

2.3.1 NATURAL–NATURAL FIBER–REINFORCED HYBRID COMPOSITES

Natural–natural fiber composites are considerable due to their biodegradability, high strength, and low cost. But there are some problems, such as their low temperature resistance, low strength, high moisture absorption, and reduction in stiffness, flexural strength, tensile strength, and impact properties due to increases in moisture absorption. Table 2.1 shows the hybrid fiber composites made from a combination of natural–natural fibers.

TABLE 2.1 Natural-Natural Fibers-Reinforced Hybrid Composites (Amir et al., 2019)

Author	Year	Hybrid Fiber	Resin	Fabrication Methods
Jacob et al.	2004	Sisal/oil palm	Hexamethylene	Compression molding process
Czigang	2006	Basalt/hemp	Polypropylene	Hot pressing
Noorunnisa et al.	2007	Sisal/silk	Polyester	Hand lay-up method
Thiruchitra et al.	2009	Banana/kenaf	Polyester	Hand lay-up method
Idicula et al.	2009	Banana/sisal	Polyester	Hand lay-up method
Jawaid et al.	2011	Oil palm/jute	Epoxy	Compression molding process
Jawaid et al.	2013	Jute/oil palm	Epoxy	Hand lay-up method
Bhoopalan et al.	2013	Jute/banana	Epoxy	Hand lay-up method
Nisini et al.	2017	Basalt/flax	Epoxy	Hand lay-up method
Cavalcanti et al.	2019	Jute/sisal/curaua	Epoxy	Hand lay-up method
Sarasini et al.	2017	Ramie/borassus	Epoxy	Injection molding process

2.3.2 SYNTHETIC–SYNTHETIC FIBER-REINFORCED HYBRID COMPOSITES

Synthetic–synthetic fibers used for hybrid fiber composites include carbon fibers, glass fibers, *Kevlar*, and so forth. Synthetic fibers are manufactured by an energy-intensive process that produces toxic by-products, which are difficult to recycle and resistant to biodegradation. As awareness of the effects of environmental pollution increases, consumers are eliminating the use of nondegradable synthetic polymer products and fibers. Table 2.2 shows the hybrid fiber composites made from a combination of synthetic–synthetic fibers.

TABLE 2.2 Synthetic–Synthetic Fiber-Reinforced Hybrid Composites (Amir et al., 2019)

Author	Year	Hybrid Fiber	Resin	Fabrication Methods
Rimdusit et al.	2012	*Kevlar*/polycarbonate/ acrylonitrile–buta-diene–styrene	Polycarbonate	Compression molding process
Al-Jeebory et al.	2010	Carbon/*Kevlar*/ araldite matrix	Epoxy	Hand lay-up method
Guru Raja et al.	2013	*Kevlar*/glass	Epoxy	Hand lay-up method
Randjbaran et al.	2014	*Kevlar*/carbon/glass	Polyparaphenylene	Hand lay-up method

2.3.3 NATURAL–SYNTHETIC FIBER-REINFORCED HYBRID COMPOSITES

Natural–synthetic hybrid composites increase temperature resistance, provide high strength, and reduce moisture absorption. Therefore, natural–synthetic hybrid fiber composites can be used on a large scale in the manufacturing industry, supported by experimental testing and investigation (Amir et al., 2019). Table 2.3 shows the hybrid fiber composites made from a combination of natural–synthetic fibers.

2.4 FABRICATION METHODS OF HYBRID FIBER COMPOSITES

There are two broad divisions of the composites manufacturing process (Balasubramanian et al., 2018; Fleischer et al., 2018; Hiremath, 2020; McIlhagger et al., 2020; Zin et al., 2016). These are:

TABLE 2.3 Natural–Synthetic Fiber-Reinforced Hybrid Composites

Author	Year	Hybrid Fiber	Resin	Fabrication Methods
Thwe and Liao	2002	Bamboo/glass	Polypropylene	Compression molding process
Mishra et al.	2003	Pineapple leaf/ sisal/glass	Polyester	Hydraulic press
Thwe and Liao	2003	Bamboo/glass	Polypropylene	Injection molding
Velmurugan and Manikandan	2007	Palmyra/glass	Rooflite	Hand lay-up method
Kumar et al.	2009	Coir/glass	Phenolic	Compression molding with hand lay-up
Govardhan and Rao	2012	*Roystonea regia*/ glass	Epoxy (L-12)	Hand lay-up method
Hemath et al.	2017	Sugarcane/glass	Epoxy	Hand lay-up method
Atiqah et al.	2018	Sugar palm/glass	Thermoplastic polyurethane	Hand lay-up method
Giridharan	2019	Ramie/glass	Epoxy	Hand lay-up method

- Open-molding methods; and
- Closed-molding methods.

2.4.1 OPEN-MOLDING METHODS

In open-molding methods, the gel coat and laminates are exposed to the atmosphere during fabrication. The various methods of open molding are hand lay-up, spray-up, autoclave curing, filament winding, and so forth. This method is used for producing vehicle cabs and fenders, boats, showers, and composite products of fiberglass. In this process, gel coat is applied to the mold surface preparation, and reinforcements are then applied by hand or in the spray-up process onto the previously gel-coated surface. Additional layers (fiberglass, resin, and catalyst) are added to improve the strength as desired. Then air is rolled out from the laminates by a hand roller, and after that the mold is left to cure. This process is flexible for all composite fabrication and is used for large-sized products that cannot be produced through an automated process (Rajak et al., 2019; Wang et al., 2011; Zin et al., 2016). Figure 2.4 shows the schematic diagram of open-molding methods.

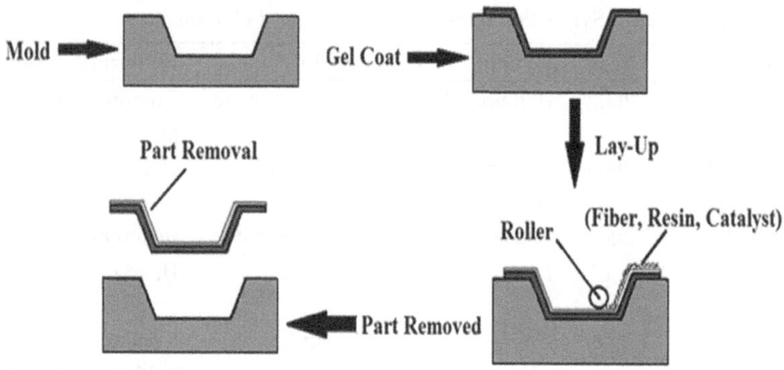

FIGURE 2.4 Open-molding methods.

Source: Adapted from Rajak et al., 2019.

2.4.1.1 HAND LAY-UP METHOD

In this process, a mold release agent is applied to the mold exterior for ease of removal of the completed composite product. The composites are prepared by placing the fiber reinforced with resin in a mold. For consolidating the composites, a roller is used, and the composites are made layer by layer. A chemical reaction initiated in the resin by a catalytic agent causes toughening of the final finished product. Hand lay-up techniques are used where the manufactured volume of composite is low. This process is used for making ducts, boat hulls, tanks, pools, furniture, sulcate, and flat sheets. The schematic diagram of the hand lay-up method is shown in Figure 2.5 (McIlhagger et al., 2020).

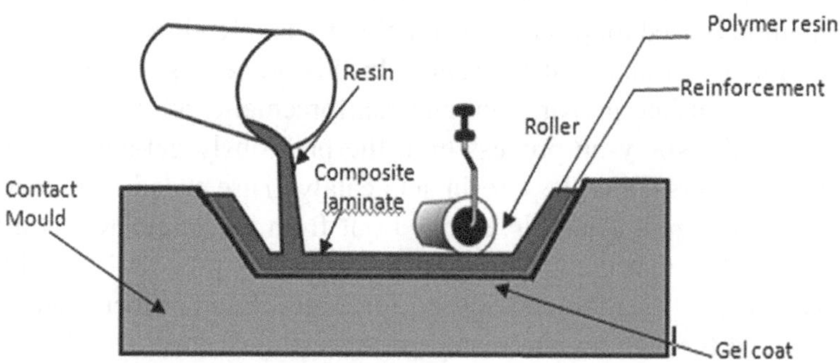

FIGURE 2.5 Hand lay-up process.

Source: Reprinted with permission from McIlhagger et al., 2015. Copyright 2015. Elsevier.

2.4.1.2 SPRAY LAY-UP METHOD

It is similar to the hand lay-up process. In this technique, the fiber is chopped in a handheld gun and fed into a spray of resin directed at the mold, and dropped materials are left to cure under atmospheric conditions. A skilled operator is required to perform the spray lay-up process. The polyester resin is used with glass roving for the spray lay-up process. It is faster and less costly compared to the hand lay-up process. It should be noted that the spray lay-up process produces short-fiber composites, whereas the hand lay-up process uses continuous woven fiber as well (Cripps et al., 2000; Middleton, 2015; Rajak et al., 2019). The spray lay-up process is shown in Figure 2.6.

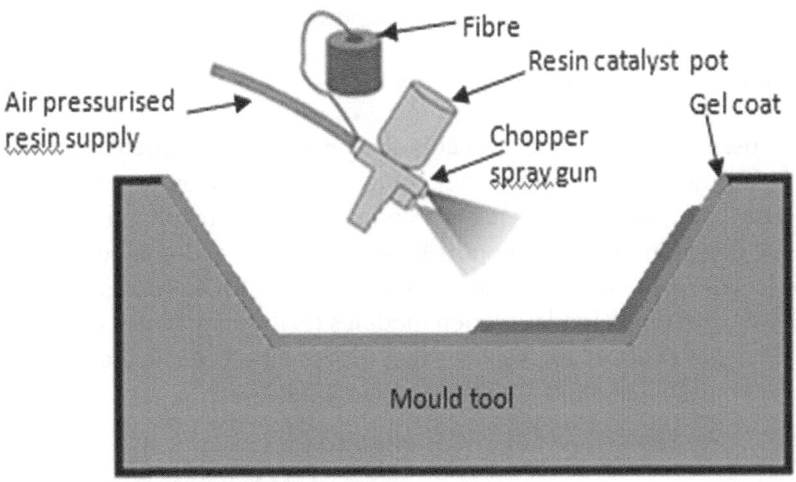

FIGURE 2.6 Spray lay-up process.

Source: Reprinted with permission from Middleton, 2015. Copyright 2015. Elsevier.

2.4.1.3 AUTOCLAVE CURING

Autoclave curing is done in a closed container to control the pressure and temperature. This process is used for curing matrix composites and is combined with the curing pressure system. The cured composites are prepared by the hand lay-up technique and have been saturated with resin. Then the composites are placed in an autoclave, and after curing, the composite is solidified (Campbell, 2004; Eckold, 1994). The autoclave curing process is shown in Figure 2.7.

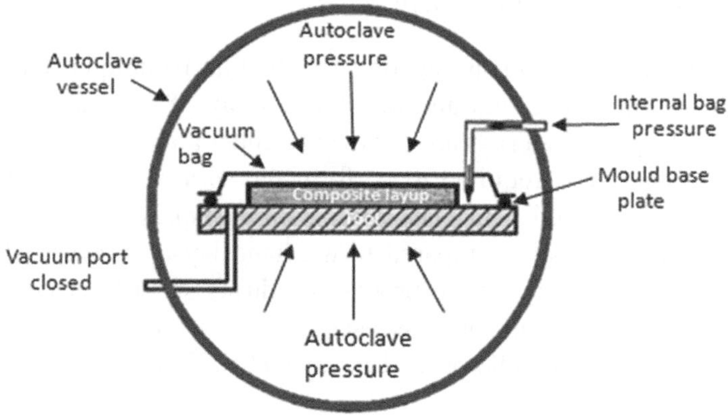

FIGURE 2.7 Autoclave curing.

Source: Reprinted with permission from Eckold, 1994. Copyright 1994. Elsevier

2.4.1.4 FILAMENT WINDING

It is used to produce components that are generally circular in shape. In this process, continuous fiber rovings are drawn by a puller through a resin bath. The composites are coiled around a revolving mandrel in the x-direction so that smooth dissemination of the composites along the length of the mandrel can be accomplished. This process is used for hollow, oval, and circular section modules (Balasubramanian et al., 2018; McIlhagger et al., 2020; Rajak et al., 2019). The schematic diagram of the closed-molding method is shown in Figure 2.8.

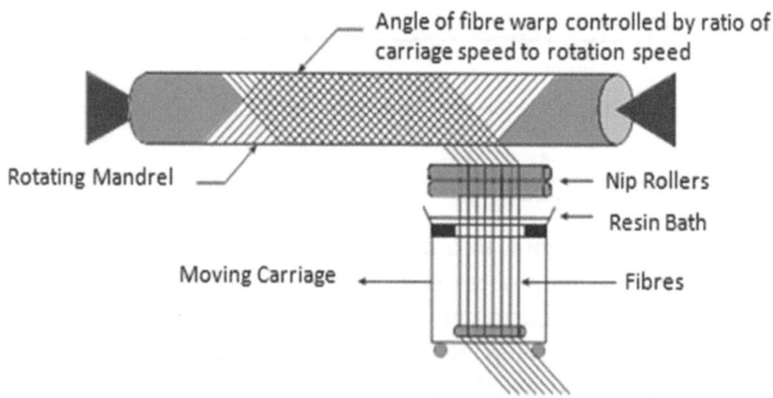

FIGURE 2.8 Filament winding.

Source: Reprinted with permission from McIlhagger et al., 2015. Copyright 2015. Elsevier.

2.4.2 CLOSED-MOLDING METHODS

In closed-molding methods, the composites are managed in a two-part mold set, with a vacuum bag, or between heated matched dies. The various methods of the closed-molding process are compression molding, pultrusion, vacuum infusion processing, vacuum bag molding, resin transfer molding, and so forth. These processes are used for long production runs and components, from small domestic items to doors, cab panels for vehicles, and so forth. The feed material flows into the contour of the mold, and the time for completing the molding process depends on the feed materials and dimensions of the components. The fabrication method for composites designated by a manufacturer depends on factors such as cost, shape, number of components, and required performance (Wang et al., 2011; Zin et al., 2016). The schematic diagram of the closed-molding method is shown in Figure 2.9.

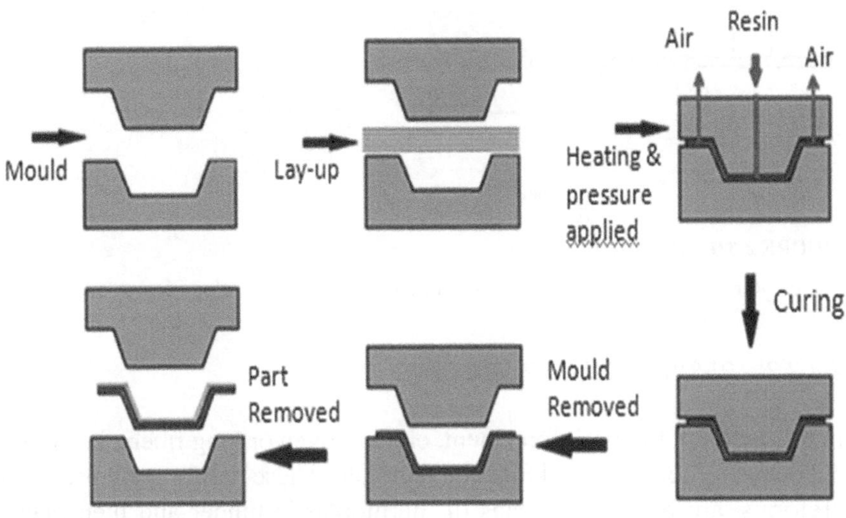

FIGURE 2.9 Closed-molding method.

Source: Adapted from Fleischer et al., 2018

2.4.2.1 PULTRUSION

Pultrusion is an automated process for manufacturing composites into continuous and cross-section profiles. In this process, the product is formed by drawing the materials instead of pushing them through the die, as in the case of extrusion. Continuous fiber rovings are drawn by means of a puller over a resin bath of thermosetting polymer. In the end, the materials are dried to their ultimate shape, and final products are cut into the required shape and size. Resins like epoxy, polyester, and phenolic can be used with continuous fiber (Hiremath, 2020; Wang et al., 2011). Figure 2.10 shows the pultrusion process used to manufacture hybrid composites.

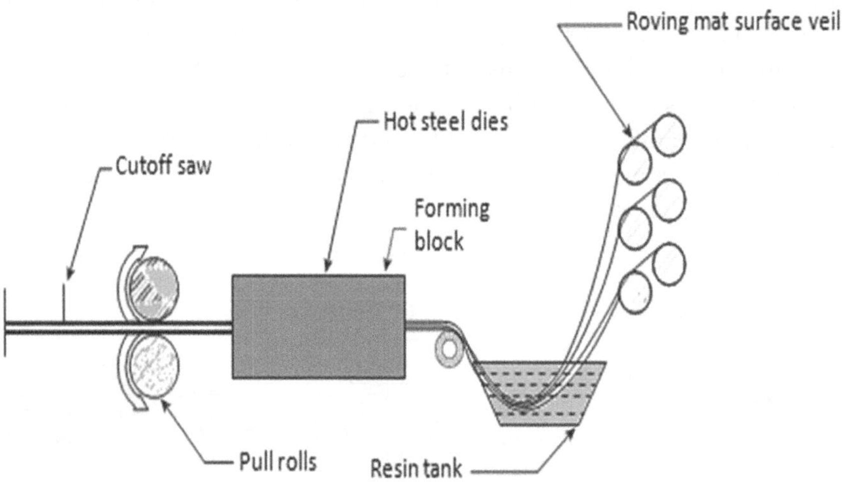

FIGURE 2.10 Pultrusion.

Source: Reprinted with permission from McIlhagger et al., 2015. Copyright 2015. Elsevier.

2.4.2.2 RESIN TRANSFER MOLDING

In this method, fiber reinforcement, either woven or long fibers, is initially cut out using scissors or a knife and template. These reinforcements, called preforms, are bound by means of thermoplastic binder and then placed inside the mold cavity in the closed-mold process. Resin is transferred into the mold cavity by means of varying equipment, such as pressure or vacuum. When the mold cavity is full, the resin supply is removed, the mold cavity inlets and outlets are sealed, and heat is applied to cure the resin (FC and Campbell Flake, 2004). The resin transfer molding process is shown in Figure 2.11.

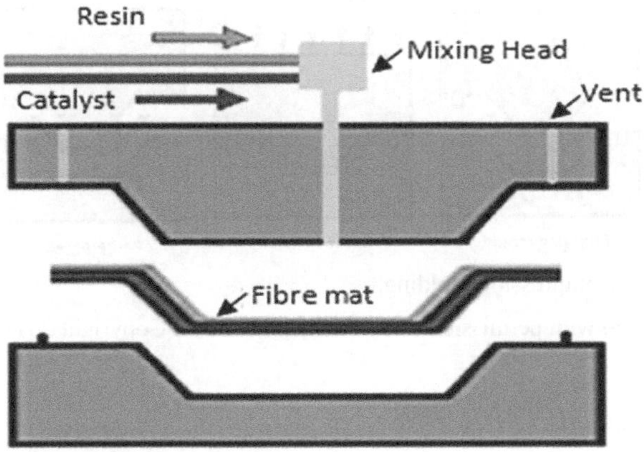

FIGURE 2.11 Resin transfer molding.

2.4.2.3 VACUUM BAGGING

Vacuum bagging is a composite manufacturing technique that uses a vacuum bag to confer pressure and consolidate layers within the laminate. It is an advanced version of the wet lay-up process. In this technique, the molds are placed horizontally, and composite prepreg laminates are arranged onto the mold base. Then a series of materials and equipment are enclosed onto the composite layers, such as an air-bleeder layer, release film layer, breather layer, and blocked film layer. In the end, a vacuum bag is placed to cover all the layers of the composite laminates. Then the vacuum bag is sealed by means of sealers (McIlhagger et al., 2020; Wang et al., 2011).

2.4.2.4 COMPRESSION MOLDING

Compression moldings are of two types: first, cold compression molding, and second, hot compression molding. In this method, for thermosetting matrices, a normal intermediate material called molding compound is generally used. It is a semi-cured composite. In cold compression molding, curing takes place at room temperature, while in hot compression molding, heat is applied to the mold and then transferred to the composites (McIlhagger et al., 2020; Wang et al., 2011). The compression molding setup is shown in Figure 2.12.

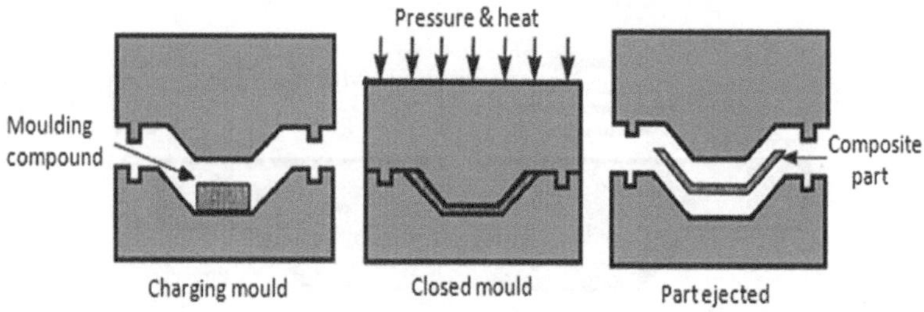

FIGURE 2.12 Compression molding.

Source: Reprinted with permission from Wang et al., 2011. Copyright 2011. Elsevier.

2.4.2.5 INJECTION MOLDING

In this technique, injection-molded compound in the form of small pellets or granules is prepared by using a twin-screw extruder. Then the granules are fed inside the mold through a heated barrel and a hopper. The shearing action of a reciprocating screw and the heating process in the heated barrel cause the injection molding compound to melt. The molten composites are injected into the mold to form a composite part. Then the composites are ejected from the mold (Liu, 2012). Figure 2.13 shows the injection molding fabrication setup for producing hybrid composites.

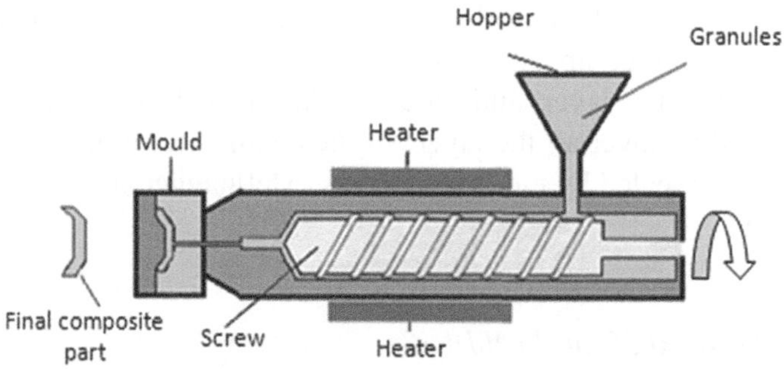

FIGURE 2.13 Injection molding.

Source: Reprinted with permission from Liu, 2012. Copyright 2012. Elsevier.

2.5 ENGINEERING APPLICATIONS OF HYBRID FIBER COMPOSITES

2.5.1 AEROSPACE INDUSTRY

The aerospace industry requires extremely lightweight materials for the manufacturing of wings with variable geometry that run on solar or renewable energy and make no noise. Natural-synthetic hybrid composites, which have high potential strength, greater performance functionality, and environmental superiority, can be used in the aerospace industry for the manufacturing of interior parts of aircraft. The use of natural-synthetic fiber composites in the aircraft industry optimizes the design of complex structures and ensures reliable performance under humid conditions at elevated temperatures. The use of natural-synthetic hybrid fiber composites in the aerospace industry reduces the weight of aircraft by replacing steel materials. It improves the elastic strain capacity and high strength-to-weight ratio as compared to steel (McIlhagger et al., 2020; Nguyen et al., 2017). Hybrid fiber composites are light in weight, low in cost, and used in the aerospace industry for minimum cost and maximum strength. This approach is developing into a wide variety of possible uses of hybrid composites in the aerospace industry. Self-cleaning and self-repairing hybrid fiber materials are used in the aerospace industry. These types of materials can be applied to the fabric of seats and carpets in aircraft. Self-healing epoxy composites protect aircraft structures from corrosion by coating them due to their protection ability after damage (Abramovich, 2017; Gururaja and Rao, 2012; Liu, 2012; Zhang et al., 2018). Figure 2.14 shows some of the applications of hybrid composites in the aerospace industry.

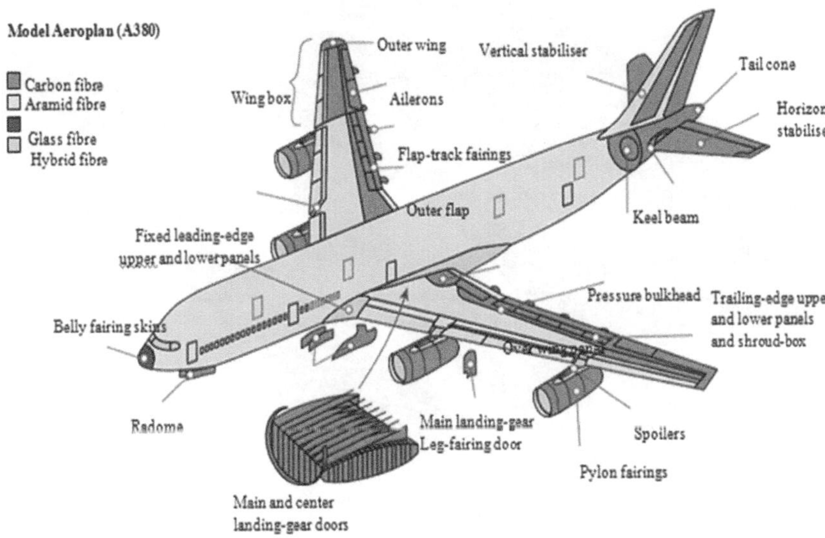

FIGURE 2.14 Application of hybrid fibers in the aerospace industry.

Source: Reprinted with permission from Abramovich, 2017. Copyright 2017. Elsevier.

2.5.2 AUTOMOTIVE INDUSTRY

Synthetic-synthetic and natural-synthetic fiber-reinforced hybrid composites have been used in making car parts like door panels, interior cabin parts, car dashboards, car bonnets, car bodies, truck fairings, vehicle floors, vehicle sheets, vehicle body parts, helmets, and so forth. The external parts of the car, such as the door panel or hood, require appropriate stiffness and offer maximum strength for dent resistance. The main reason for using natural fiber in hybrid composites in the automotive field is because it is biodegradable, has higher strength, and reduces the vehicle's weight (Badie et al., 2011; Friedrich and Almajid, 2013). Major automotive manufacturers such as BMW, Audi, Rover, and Volvo have used biofiber for making various components of their vehicles. The main components made of hybrid fibers include the door panel, seat back, spare tire lining, windshield, pillar cover panel, storage panel, exterior underbody protection trim, and so forth (Friedrich and Almajid, 2013; Kim and Chalivendra, 2020). Figure 2.15 shows the composites used in the structural parts of automobiles.

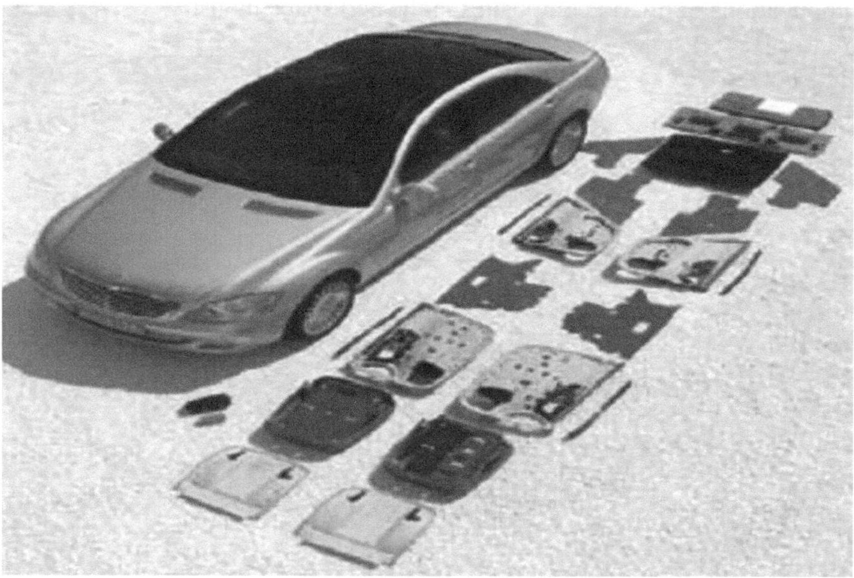

FIGURE 2.15 Application of hybrid fibers in automotives/Mercedes.

Source: Reprinted with permission from Kim, 2012. Copyright 2012. Elsevier.

2.5.3 MARINE TRANSPORTATION

Synthetic-synthetic and natural-synthetic fiber hybrid composites are used for a wide range of naval structures, including the superstructure of vessels, propellers, and aircraft carriers. Hybrid glass fiber reduces the hull weight for the construction of fast patrol boats and hydrofoils. Power boats are constructed of hybrid composites. Marine applications include the superstructure of ships, warships, sonar domes, oil and water storage tanks, and so forth. Marine applications also include ladders, deck drains, masts, and stacks. Hybrid fiber composites have higher strength, a low weight ratio, and a lower requirement for maintenance to form complex hull shapes (Mouritz et al., 2001; Pellicer et al., 2017). Figure 2.16 shows one of the applications of hybrid composites in ships.

FIGURE 2.16 Application of hybrid composites in marine transportation.

Source: Reprinted with permission from Mouritz et al., 2001. Copyright 2001. Elsevier.

2.5.4 CIVIL INFRASTRUCTURE

Hybrid composite materials are good alternatives to construction materials in various structural applications due to their high strength, high stiffness, excellent corrosion resistance, and excellent durability. The use of carbon fiber and glass fiber in girders has exploited their lightweight and excellent mechanical properties at a low cost. The use of carbon fiber and glass fiber in the flanges has maximized structural performance while reducing the overall cost. The use of synthetic-synthetic hybrid fiber composites in beams maximizes the strength of the beam, replacing existing deficient bridges and reducing weight and overall cost (Aboelseoud and Myers, 2019; Hai et al., 2010; McConnell, 2012). Figure 2.17 shows one of the applications of hybrid composites in civil infrastructure.

FIGURE 2.17 Application of hybrid FRP in civil infrastructure.

Source: Reprinted with permission from McConnell, 2012. Copyright 2012. Elsevier.

2.5.5 SMART STRUCTURE

Carbon and glass fibers are used in intelligent structures, functional structures, and adaptive structures. They mainly consist of four key elements: actuators, sensors, control strategies, and power conditioning electronics. Glass-*Kevlar* fibers are used for smart structures because of their distinctive characteristics such as low cost, light weight, ease of manufacture, biodegradability, and high shape deformability as SMP composites. Fiber-reinforced polymer bars replace conventional steel bars in concrete structures. Hybrid fiber sheets are used for columns, column-beam joints, and masonry work (Chan and Zhou, 2014; Nguyen et al., 2017).

2.5.6 SPORTS

Hybrid fiber composites are used in the sports industry due to their light weight, high specific strength, and excellent durability. Hybrid fibers (glass/carbon/*Kevlar*/flax/hemp) are used to produce helmets, bicycle frames, aerobars of bicycles, tennis rackets, snowboards, and fishing rods, which feature light weight and anti-vibration characteristics (McConnell, 2008). Figure 2.18 shows the various parts of sporting goods.

FIGURE 2.18 Application of hybrid fibers in sporting goods.

Source: Reprinted with permission from McConnell, 2008. Copyright 2008. Elsevier.

2.5.7 ENERGY SECTORS

Wind power is a renewable energy source that is essentially inexhaustible and a clean fuel source. Conventional energy sources produce greenhouse gases, which increase the average temperature of the world. Wind turbines do not produce greenhouse gases and are typically made from glass fibers reinforced with epoxy composites or polyester matrix composites. Synthetic fiber-reinforced turbine blades have limited applications due to their high cost and restricted manufacturing conditions (Ciang et al., 2008; Marsh, 2014). Figure 2.19 shows the use of composites in windmills.

FIGURE 2.19 Application of hybrid fibers in energy sector components.

Source: Reprinted with permission from Marsh, 2014. Copyright 2014. Elsevier.

2.5.8 STRUCTURAL CONCRETE COMPOSITES

Hybrid fiber composites are widely accepted in the construction industry due to their good resistance to corrosion and high strength-to-weight ratio. Concrete-filled hybrid fiber tubes are used as poles, piling, highway overhead sign/symbol structures, and components of bridges. Hybrid fiber-reinforced composites have good compressive strength, good tensile strength, and are low in cost (Hai et al., 2010; Xie et al., 2018; Zou et al., 2016). Figure 2.20 shows the perforated FRP rib for the FRP-concrete hybrid beam/deck.

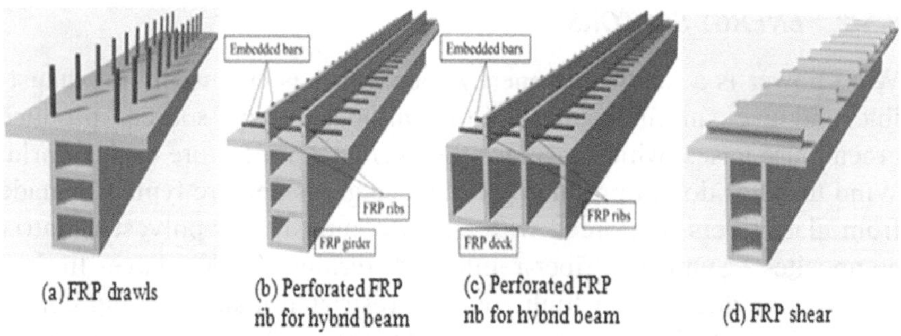

FIGURE 2.20 Application of perforated hybrid FRP rib for FRP-concrete hybrid beam/deck.

Source: Reprinted with permission from Zou et al, 2016. Copyright 2016. Elsevier.

2.5.9 *TELECOMMUNICATION COMPONENTS*

The growth in the telecommunication industry is increasing the demand for high-speed data transmission with higher power capacity. Fiber optic technology has a large number of applications in fiber optic communication for high-speed communication. Fiber optics are flexible, thin, and as fine as human hair. Optical fibers are used in submarines as a key element for communication technology under the sea. Carbon nanotubes (CNTs), which are based on hybrid fibers, have high electrical and thermal conductivity to improve electrochemical properties and to enhance energy conversion and storage capacities of laser-induced breakdown spectroscopy (Chan and Zhou, 2014).

2.6 CONCLUSION

The hybridization of natural fiber composites is one of the most efficient ways to improve the mechanical properties of composite ingredients. Natural-synthetic fiber-based hybrid composites create a competitive market for industrial and automotive applications. This analysis is carried out to study the effect of hybrid composites on enhancements in thermal and mechanical properties for different engineering applications. This article evaluates natural and synthetic fiber-reinforced hybrid composites, which have high strength, low density, and are less expensive compared to synthetic composites. Therefore, they offer better utilization in commercial sectors like the automotive industry, building construction, aerospace, commercial airlines, construction, and electronic product industries. Thermosetting resins such as polyester, epoxy, phenolic, polyamide, and phenolic are generally used with matrix materials. Carbon-, glass-, and *Kevlar*-reinforced hybrid composites are used in the construction industry. Advancements in hybrid fibers have helped manufacturing technology reduce costs while expanding their applications.

KEYWORDS

- **aerospace**
- **automotive industry**
- **energy sector**
- **hybrid composites**
- **hybridization**
- **marine transportation**
- **natural fiber**

REFERENCES

Aboelseoud, M. A., & Myers, J. J. (2019). Flexural Behavior of Hybrid Composite Beam (HCB) Bridges. *Advances in Materials Science and Engineering, 2019.*

Abramovich, H. (2017). Introduction to composite materials. In *Stability and Vibrations of Thin Walled Composite Structures* (pp. 1–47). Woodhead Publishing.

Al-Jeebory, A., & Al-Mosawi, A. (2010). Mechanical properties of araldite matrix composites reinforced with hybrid carbon-Kevlar fibers. *Al-Qadisiya Journal for Engineering Sciences.*

Amir, S. M. M., Sultan, M. T. H., Jawaid, M., Ariffin, A. H., Mohd, S., Salleh, K. A. M., ... & Shah, A. U. M. (2019). Nondestructive testing method for Kevlar and natural fiber and their hybrid composites. In *Durability and life prediction in biocomposites, fiber-reinforced composites and hybrid composites* (pp. 367–388). Woodhead Publishing. doi: 10.1016/B978-0-08-102290-0.00016-7.

Assarar, M., Zouari, W., Sabhi, H., Ayad, R., & Berthelot, J. M. (2015). Evaluation of the damping of hybrid carbon–flax reinforced composites. *Composite Structures, 132*, 148–154. doi: 10.1016/j.compstruct.2015.05.016.

Atiqah, A., Jawaid, M., Sapuan, S. M., Ishak, M. R., & Alothman, O. Y. (2018). Thermal properties of sugar palm/glass fiber reinforced thermoplastic polyurethane hybrid composites. *Composite Structures, 202*, 954–958. doi: 10.1016/j.compstruct.2018.05.009.

Badie, M. A., Mahdi, E., & Hamouda, A. M. S. (2011). An investigation into hybrid carbon/glass fiber reinforced epoxy composite automotive drive shaft. *Materials & Design, 32*(3), 1485–1500. doi: 10.1016/j.matdes.2010.08.042.

Balasubramanian, K., Sultan, M. T., & Rajeswari, N. (2018). Manufacturing techniques of composites for aerospace applications. In *Sustainable Composites for Aerospace Applications* (pp. 55–67). Woodhead Publishing. doi: 10.1016/B978-0-08-102131-6.00004-9.

Bandaru, A. K., Patel, S., Sachan, Y., Ahmad, S., Alagirusamy, R., & Bhatnagar, N. (2016). Mechanical behavior of Kevlar/basalt reinforced polypropylene composites. *Composites Part A: Applied Science and Manufacturing, 90*, 642–652.

Begum, K., & Islam, M. (2013). Natural fiber as a substitute to synthetic fiber in polymer composites: a review. *Res. J. Eng. Sci, 2278*, 9472.

Boopalan, M., Niranjanaa, M., & Umapathy, M. J. (2013). Study on the mechanical properties and thermal properties of jute and banana fiber reinforced epoxy hybrid composites. *Composites Part B: Engineering, 51*, 54–57. doi: 10.1016/j.compositesb.2013.02.033.

Campbell Flake, C. (2004). Chapter 6: Curing: It's a Matter of Time (t), Temperature (T) and Pressure (P). *1*, 175–221. doi: 10.1016/j.compstruct.2018.05.021.

Campbell, F. C. (2004). In: *Curing: It's a Matter of Time (t), Temperature (T), and Pressure (P)* (pp. 175–221).

Cavalcanti, D. K. K., Banea, M. D., Neto, J. S. S., Lima, R. A. A., Da Silva, L. F. M., & Carbas, R. J. C. (2019). Mechanical characterization of intralaminar natural fiber-reinforced hybrid composites. *Composites Part B: Engineering, 175*, 107149. doi: 10.1016/j.compositesb.2019.107149.

Chan, Y. W. S., & Zhou, Z. (2014). Advances of FRP-based smart components and structures. *Pacific Science Review, 16*(1), 1–7.

Ciang, C. C., Lee, J. R., & Bang, H J. (2008). Structural health monitoring for a wind turbine system: a review of damage detection methods. *Measurement Science and Technology*, *19*(12), 122001. doi: 10.1088/0957-0233/19/12/122001.

Cripps, D., Searle, T. J., & Summerscales, J. (2000). Open mold techniques for thermoset composites. doi: 10.1016/b0-08-042993-9/00188-1.

Czél, G., & Wisnom, M. R. (2013). Demonstration of pseudo-ductility in high performance glass/epoxy composites by hybridization with thin-ply carbon prepreg. *Composites Part A: Applied Science and Manufacturing*, *52*, 23–30. doi: 10.1016/j.compositesa.2013.04.006.

Czigány, T. (2006). Special manufacturing and characteristics of basalt fiber reinforced hybrid polypropylene composites: mechanical properties and acoustic emission study. *Composites Science and Technology*, *66*(16), 3210–3220. doi: 10.1016/j.compscitech.2005.07.007.

De Rosa, I. M., Iannoni, A., Kenny, J. M., Puglia, D., Santulli, C., Sarasini, F., & Terenzi, A. (2011). Poly (lactic acid)/*Phormium tenax* composites: Morphology and thermo-mechanical behavior. *Polymer Composites*, *32*(9), 1362–1368. doi: 10.1002/pc.

Eckold, G. (1994). "Manufacture," in *Manufacture. Design and Manufacture of Composite Structures*, pp. 251–304.

Evans, J. T. (1992). Analysis and Performance of Fiber Composites. By Bhagwan D. Agarwal and Lawrence J. Broutman; Published by John Wiley and Sons Inc. doi: 10.1016/0921-5093(92)90189-8.

Fleischer, J., Teti, R., Lanza, G., Mativenga, P., Möhring, H. C., & Caggiano, A. (2018). Composite materials parts manufacturing. *CIRP Annals*, *67*(2), 603–626. doi: 10.1016/j.cirp.2018.05.005.

Friedrich, K., & Almajid, A. A. (2013). Manufacturing aspects of advanced polymer composites for automotive applications. *Applied Composite Materials*, *20*, 107–128. doi: 10.1007/s10443-012-9258-7.

Giridharan, R. (2019). Preparation and property evaluation of Glass/Ramie fibers reinforced epoxy hybrid composites. *Composites Part B: Engineering*, *167*, 342–345.

Goud, G., & Rao, R. N. (2012). Mechanical and electrical performance of Roystonea regia/glass fiber reinforced epoxy hybrid composites. *Bulletin of Materials Science*, *35*, 595–599. doi: 10.1007/s12034-012-0324-4.

Guru Raja, M. N., & Hari Rao, A. N. (2013). Effect of an angle-ply orientation on tensile properties of Kevlar/glass hybrid composites. *International Journal on Theoretical and Applied Research in Mechanical Engineering*, *2*, 63–67.

Gururaja, M. N., & Rao, A. H. (2012). A review on recent applications and future prospectus of hybrid composites. *International Journal of Soft Computing and Engineering*, *1*(6), 352–355.

Hai, N. D., Mutsuyoshi, H., Asamoto, S., & Matsui, T. (2010). Structural behavior of hybrid FRP composite I-beam. *Construction and Building Materials*, *24*(6), 956–969. doi: 10.1016/j.conbuildmat.2009.11.022.

Hiremath, S. S. (2020). *Natural Fiber Reinforced Composites in the Context of Biodegradability: A Review*. doi: 10.1016/b978-0-12-803581-8.11418-3.

Ho, M. P., Wang, H., Lee, J. H., Ho, C. K., Lau, K. T., Leng, J., & Hui, D. (2012). Critical factors on manufacturing processes of natural fiber composites. *Composites Part B: Engineering*, *43*(8), 3549–3562. doi: 10.1016/j.compositesb.2011.10.001.

Idicula, M., Joseph, K., & Thomas, S. (2010). Mechanical performance of short banana/sisal hybrid fiber reinforced polyester composites. *Journal of Reinforced Plastics and Composites*, *29*(1), 12–29. doi: 10.1177/0731684408095033.

Jacob, M., Thomas, S., & Varughese, K. T. (2004). Mechanical properties of sisal/oil palm hybrid fiber reinforced natural rubber composites. *Composites Science and Technology*, *64*(7-8), 955–965. doi: 10.1016/S0266-3538(03)00261-6.

Jawaid, M. H. P. S., & Khalil, H. A. (2011). Cellulosic/synthetic fiber reinforced polymer hybrid composites: A review. *Carbohydrate Polymers*, *86*(1), 1–18. doi: 10.1016/j.carbpol.2011.04.043.

Jawaid, M., Khalil, H. A., Hassan, A., Dungani, R., & Hadiyane, A. (2013). Effect of jute fiber loading on tensile and dynamic mechanical properties of oil palm epoxy composites. *Composites Part B: Engineering*, *45*(1), 619–624. doi: 10.1016/j.compositesb.2012.04.068.

Kim, Y. K., & Chalivendra, V. (2020). Natural fiber composites (NFCs) for construction and automotive industries. In *Handbook of Natural fibers* (pp. 469–498). Woodhead Publishing.

Kumar, C. S., Arumugam, V., Dhakal, H. N., & John, R. (2015). Effect of temperature and hybridisation on the low velocity impact behavior of hemp-basalt/epoxy composites. *Composite Structures*, *125*, 407–416., doi: 10.1016/j.compstruct.2015.01.037.

Kumar, G. H., Babuvishwanath, H., Purohit, R., Sahu, P., & Rana, R. S. (2017). Investigations on mechanical properties of glass and sugarcane fiber polymer matrix composites. *Materials Today: Proceedings*, *4*(4), 5408–5420. doi: 10.1016/j.matpr.2017.05.052.

Kumar, N. M., Reddy, G. V., Naidu, S. V., Rani, T. S., & Subha, M. C. S. (2009). Mechanical properties of coir/glass fiber phenolic resin based composites. *Journal of Reinforced Plastics and Composites*, *28*(21), 2605–2613. doi: 10.1177/0731684408093092.

Liu, S. J. (2012). Injection molding in polymer matrix composites. In *Manufacturing Techniques for Polymer Matrix Composites (PMCs)* (pp. 15–46). Woodhead Publishing. doi: 10.1016/B978-0-85709-067-6.50002-X.

Marsh, G. (2014). Greater role for composites in wind energy. *Reinforced Plastics*, *58*(1), 20–24. doi: 10.1016/S0034-3617(14)70037-2.

McConnell, V. P. (2008). Advanced performance elevates sporting spirit. *Reinforced Plastics*, *52*(9), 24–29. doi: 10.1016/S0034-3617(08)70336-9.

McConnell, V. P. (2012). Infrastructure innovations. *Reinforced Plastics*, *56*(3), 36–42.

McIlhagger, A., Archer, E., & McIlhagger, R. (2020). Manufacturing processes for composite materials and components for aerospace applications. In *Polymer Composites in the Aerospace Industry* (pp. 59–81). Woodhead Publishing. doi: 10.1016/B978-0-85709-523-7.00003-7.

Middleton, B. (2015). Composites: manufacture and application. *Design and manufacture of Plastic Components for Multifunctionality: Structural Composites, Injection Molding, and 3D Printing*, 53.

Mishra, S., Mohanty, A. K., Drzal, L. T., Misra, M., Parija, S., Nayak, S. K., & Tripathy, S. S. (2003). Studies on mechanical performance of biofiber/glass reinforced polyester hybrid composites. *Composites Science and Technology*, *63*(10), 1377–1385. doi: 10.1016/S0266-3538(03)00084-8.

Mouritz, A. P., Gellert, E., Burchill, P., & Challis, K. (2001). Review of advanced composite structures for naval ships and submarines. *Composite Structures*, *53*(1), 21–42. doi: 10.1016/S0263-8223(00)00175-6.

Nguyen, H., Zatar, W., & Mutsuyoshi, H. (2017). Hybrid polymer composites for structural applications. *Hybrid Polymer Composite Materials*, 35–51. doi: 10.1016/B978-0-08-100785-3.00002-4.

Noorunnisa Khanam, P., Mohan Reddy, M., Raghu, K., John, K., & Venkata Naidu, S. (2007). Tensile, flexural and compressive properties of sisal/silk hybrid composites. *Journal of Reinforced Plastics and Composites*, *26*(10), 1065–1070. doi: 10.1177/0731684407079347.

Pellicer, E., Nikolic, D., Sort, J., Baró, M., Zivic, F., Grujovic, N,…. & Pelemis, S. (Eds.). (2017). *Advances in applications of industrial biomaterials* (pp. 1-214). New York, NY, USA:: Springer International Publishing. doi: 10.1007/978-3-319-62767-0.

Pickering, K. L., Efendy, M. A., & Le, T. M. (2016). A review of recent developments in natural fiber composites and their mechanical performance. *Composites Part A: Applied Science and Manufacturing*, *83*, 98–112. doi: 10.1016/j.compositesa.2015.08.038.

Rajak, D. K., Pagar, D. D., Kumar, R., & Pruncu, C. I. (2019). Recent progress of reinforcement materials: A comprehensive overview of composite materials. *Journal of Materials Research and Technology*, *8*(6), 6354–6374. doi: 10.1016/j.jmrt.2019.09.068.

Randjbaran, E., Zahari, R., Abdul Jalil, N. A., & Abang Abdul Majid, D. L. (2014). Hybrid composite laminates reinforced with Kevlar/carbon/glass woven fabrics for ballistic impact testing. *The Scientific World Journal*, *2014*. doi: 10.1155/2014/413753.

Ravishankar, B., Nayak, S. K., & Kader, M. A. (2019). Hybrid composites for automotive applications–A review. *Journal of Reinforced Plastics and Composites*, *38*(18), 835–845. doi: 10.1177/0731684419849708.

Rimdusit, S., Lorjai, P., Sujirote, K., & Tiptipakorn, S. (2012). Physical and mechanical characteristics of Kevlar fiber-reinforced PC/ABS composites. *Engineering Journal*, *16*(1), 57–66. doi: 10.4186/ej.2012.16.1.57.

Sanjay, M. R., Madhu, P., Jawaid, M., Senthamaraikannan, P., Senthil, S., & Pradeep, S. (2018). Characterization and properties of natural fiber polymer composites: A comprehensive review. *Journal of Cleaner Production*, *172*, 566–581. doi: 10.1016/j.jclepro.2017.10.101.

Sarasini, F., Tirillò, J., Puglia, D., Dominici, F., Santulli, C., Boimau, K., ... & Torre, L. (2017). Biodegradable polycaprolactone-based composites reinforced with ramie and borassusfibers. *Composite Structures*, *167*, 20–29.

Sathishkumar, T. P., Naveen, J. A., & Satheeshkumar, S. (2014). Hybrid fiber reinforced polymer composites–a review. *Journal of Reinforced Plastics and Composites*, *33*(5), 454–471. doi: 10.1177/0731684413516393.

Shalwan, A., & Yousif, B. F. (2013). In state of art: mechanical and tribological behavior of polymeric composites based on natural fibers. *Materials & Design*, *48*, 14–24. doi: https://doi.org/10.1016/j.matdes.2012.07.014.

Swolfs, Y., Gorbatikh, L., & Verpoest, I. (2014). fiber hybridization in polymer composites: A review. *Composites Part A: Applied Science and Manufacturing*, *67*, 181–200, doi: 10.1016/j.compositesa.2014.08.027.

Thiruchitrambalam, M., Alavudeen, A., Athijayamani, A., Venkateshwaran, N., & Perumal, A. E. (2009). Improving mechanical properties of banana/kenaf polyester hybrid composites using sodium lauryl sulfate treatment. *Mater. Phys. Mech*, *8*(2), 165–173.

Thwe, M. M., & Liao, K. (2002). Effects of environmental aging on the mechanical properties of bamboo–glass fiber reinforced polymer matrix hybrid composites. *Composites Part A: Applied Science and Manufacturing*, *33*(1), 43–52. doi: 10.1016/S1359-835X(01)00071-9.

Thwe, M. M., & Liao, K. (2003). Durability of bamboo-glass fiber reinforced polymer matrix hybrid composites. *Composites Science and Technology*, *63*(3–4), 375–387. doi: 10.1016/S0266-3538(02)00225-7.

Turra, A., & Baruque-Ramos, J. (2017). Corrigendum to "Synthetic fibers as microplastics in the marine environment: A review from textile perspective with a focus on domestic washings." *The Science of the Total Environment*, *603*, 836–836. doi: 10.1016/j.scitotenv.2017.06.179.

Velmurugan, R., & Manikandan, V. (2007). Mechanical properties of palmyra/glass fiber hybrid composites. *Composites Part A: Applied Science and Manufacturing*, *38*(10), 2216–2226. doi: 10.1016/j.compositesa.2007.06.006.

Wang, R. M., Zheng, S. R., & Zheng, Y. P. (2011). Introduction to polymer matrix composites. *Polym. Matrix Compos. Technol.* doi: 10.1533/9780857092229.2.253.

Xie, X., Liu, J., Li, T., Song, Y., & Wang, F. (2018). Post-formation copper-nitrogen species on carbon black: their chemical structures and active sites for oxygen reduction reaction. *Chemistry–A European Journal*, *24*(39), 9968–9975. doi: 10.1002/chem.201801764.

Zhang, X., Chen, Y., & Hu, J. (2018). Recent advances in the development of aerospace materials. *Progress in Aerospace Sciences*, *97*, 22–34. doi: 10.1016/j.paerosci.2018.01.001.

Zin, M. H., Razzi, M. F., Othman, S., Liew, K., Abdan, K., & Mazlan, N. (2016, October). A review on the fabrication method of bio-sourced hybrid composites for aerospace and automotive applications. In *IOP Conference Series: Materials Science and Engineering* (Vol. 152, No. 1, p. 012041). IOP Publishing. doi: 10.1088/1757-899X/152/1/012041.

Zou, X., Feng, P., & Wang, J. (2016). Perforated FRP ribs for shear connecting of FRP-concrete hybrid beams/decks. *Composite Structures*, *152*, 267–276. doi: 10.1016/j.compstruct.2016.05.039.

CHAPTER 3

Mechanical and Optical Properties of Side-Emitting Optical Fibers

MAVUSO AYANDA FAITH and ANUJ SHUKLA*

Durban University of Technology, Durban, KwaZulu-Natal, South Africa

*Corresponding author

ABSTRACT

Optical fibers offer a wide range of applications, including the transfer of light from a "source to a device," sending "optically encoded data," and even serving as a "temperature or strain sensor." The present work is based on a detailed study of optical fibers, their dyeing principles and methods, and experimental investigation of fiber dyeing, measurement of side emission, and plastic optical fiber cross-section analysis.

3.1 INTRODUCTION

Most fibers are used to direct light, and because optical fibers can confine light inside the fiber, they are very useful. The fiber accomplishes this by internal reflection of light in the core of the fiber along its axis. Moreover, three types of optical fibers are frequently used in telecommunications, whose type is determined by their use and end application (Zubia and Arrue, 2001). The three types are further subdivided into two categories: "single-mode fibers and multi-mode fibers," respectively.

- Multi-mode fibers with a step index;
- Multi-mode fibers with a graded index;
- Single-mode fibers.

Single-mode fibers, also known as "mono-mode fibers," have just "one mode of transmission," that is, the "light or signal wave" will go in one

Advances in Industrial Manufacturing and Materials: Integrating a New Era of Computational Modeling and Machine Learning. Debesh Mishra and Swagatika Mishra (Eds.)

DOI: 10.1201/9781779640475-3

direction only, which is linear. Because of their tiny diameter, these fibers necessitate the use of a light source with a limited "spectral breadth." These fibers are utilized when a signal must be sent at a "greater rate" and across "greater ranges." The capacity to transmit across larger distances is due to the "single light wave and compact core," which removes most signal distortion. Single-mode fibers typically transmit wavelengths ranging from "1310 to 1550 nm."

Multi-mode fibers typically have a "diameter range of 50–100 μm" as compared to single-mode fibers. The functions are generally the same—for light transmission that is transformed into a signal—however, multi-mode fibers can process "more light at faster rates." Because light waves are divided into several channels, lengthy spans of multi-mode fibers are not practical owing to interferences inside the line. Another important issue is that multi-mode fibers do not require a "narrow spectral-width light source" and hence may carry light at "shorter wavelengths (850 nm)" than single-mode fibers.

A single optical fiber is composed of multiple components, including the "core; cladding; coating buffers; strength members; and outer jackets."

The core carries the light, while the cladding aids in total "internal reflection." This is accomplished by employing cladding with a lower "refractive index" than the core, causing light transmitted in the core to "reflect off the cladding" and "remain within the core" (Figure 3.1). The refractive index is defined as the ratio of "light velocity in vacuum to light velocity in the substance." As a result, the greater the refractive index, the slower light travels through the material.

$$\textit{Index of Refraction} = \frac{\textit{Light velocity}_{vacuum}}{\textit{Light velocity}_{material}}$$

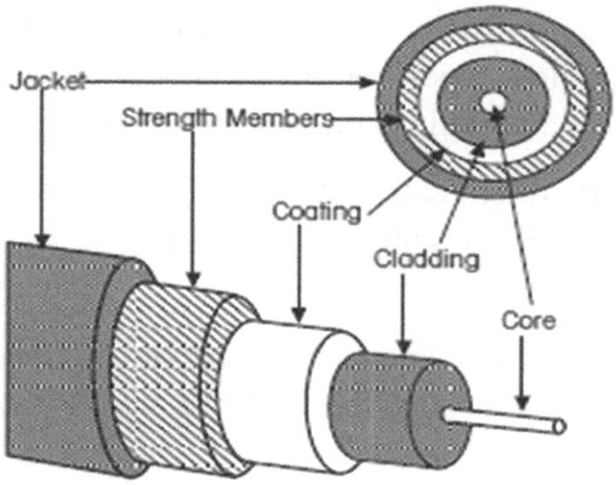

FIGURE 3.1 Construction of fiber.

3.1.1 TOTAL INTERNAL REFLECTIONS

Optical fibers work effectively because they can retain "light within the fiber core." When light moves from one substance of "differing density to another," its path "bends." At a certain point, or rather angle, "light will not penetrate through the surface but will instead bounce off it." This effect is used by optical fibers to "bend light at its border," which is the "cladding," and confine the light "within the core." Thus, by choosing material variations, such as differing refractive indices between the "core and cladding," the angle at which total internal reflection occurs may be chosen (Figure 3.2).

3.1.2 LIGHT REFLECTION FROM OPTICAL SURFACES

Light striking "an interface between two transparent optical media," such as "air and glass" or "water and glass," causes one of four things to happen to the "incident light":

- It can be "mirrored partially or completely at the interface."
- At the interface, it can be "dispersed in random directions."
- It can be "partially transmitted via the interface" and "enter the second medium," or it can be "partially absorbed in either medium."

Because the surfaces of "optical fibers" are smooth, the "rough surfaces" may be ignored.

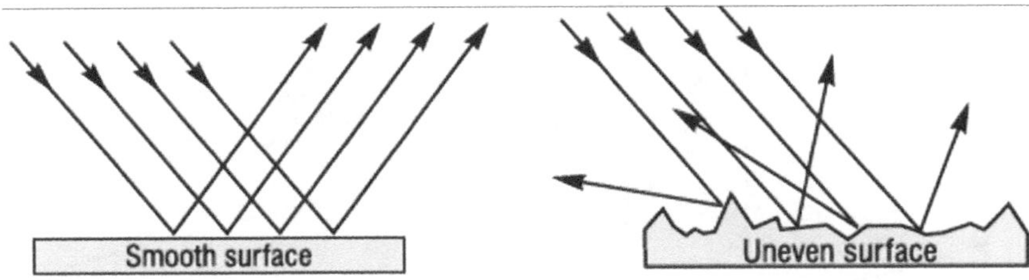

FIGURE 3.2 Specular and diffuse "reflection."

3.1.3 THE LAWS OF REFLECTION ON PLANE SURFACES

The reflection of light from "planar surfaces" (Figure 3.2) forms an angle between the reflected ray and the normal—that is, perpendicular to the surface—at the point of incidence, which is always "the same as the angle formed by the incident ray with the normal." It is worth noting that the "incident ray; reflected ray; and normal" all lie in the same plane (Figure 3.3).

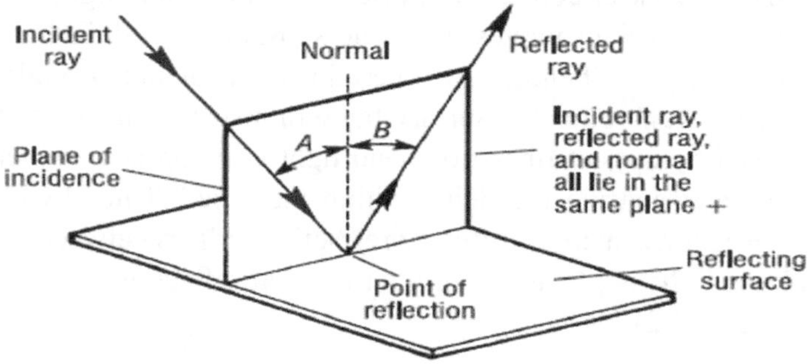

FIGURE 3.3 Reflections on plane surfaces.

3.1.4 LIGHT REFRACTIONS FROM "OPTICAL INTERFACES"

Light striking an interface, which is the geometrical plane dividing one optical medium from another, is usually partially reflected and partially transmitted. Figure 3.4 depicts a three-dimensional perspective of incident light on a partly reflecting surface or interface, which gets reflected there and refracted into the second medium. Refraction is the bending of light beams at a contact between two optical media.

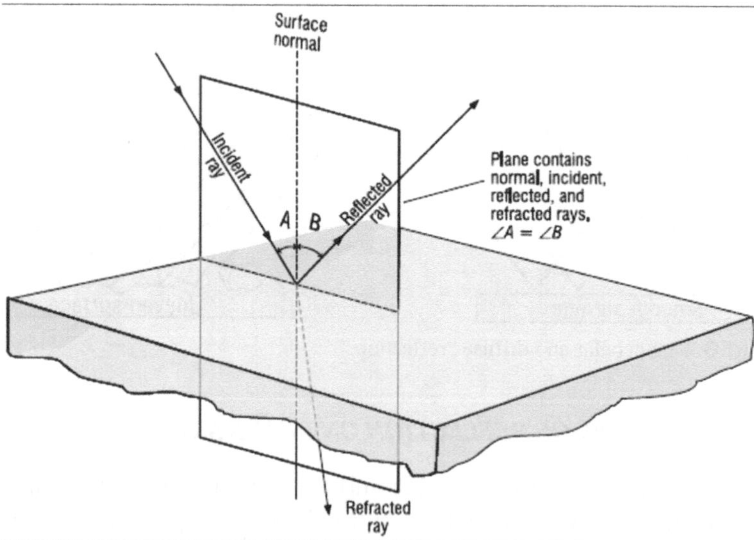

FIGURE 3.4 Light reflection and refraction at an interface.

3.1.5 REFRACTIVE INDEX

The two clear optical media comprising an interface are differentiated from one another by the "refractive index," which is a constant and denoted by the symbol n. The index of refraction of any clear optical medium is defined as the ratio of light speed in a vacuum to light speed in the medium, as indicated in the equation:

$$n = \frac{c}{v}$$

where; c is the light speed in vacuum (free space); v is the light speed in the medium; and n is the refractive index of the medium.

The refractive index in open space is one. Because it is relatively close to one for air and most gases, it is assumed to be 1.0 in most computations. It has a greater value than one for other materials (Table 3.1).

TABLE 3.1 Refractive Indexes for Common Materials at 589 nm

Common Materials	Value of n
Glass (flint)	1.66
Quartz (fused)	1.46
Sodium chloride	1.54
Water	1.33
Glycerin	1.47
Ice	1.31
Gallium arsenide	3.40
Air	1.0003
Diamond	2.42
Benzene	1.50
Ethyl alcohol	1.36
Carbon disulfide	1.63
Corn syrup	2.21
Polystyrene	1.49

The larger a medium's index of refraction, the slower light travels, and more bending occurs for the light when passing from air into the medium. Figure 3.5 depicts two general examples: one for light passing from a medium of lower index to a medium of higher index, and the other for light moving from a medium of higher index to a medium of lower index. The light beam is bent toward the normal in the first instance (low to high), while it is bent away from the normal in the second instance (high to low).

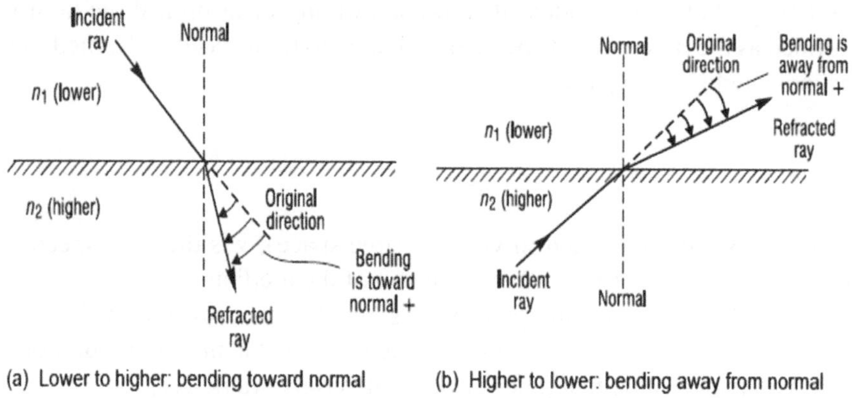

FIGURE 3.5 Refractions at the interface of media with n_1 and n_2 refractive indices.

3.1.6 *SNELL'S LAW*

The sines of the incidence angle and refraction angle at an interface between two optical media are related to the indices of refraction of the two media by Snell's equation of refraction. The rule is named after Willebrord Snell, a Dutch astronomer who developed it in the 17th century. By knowing the refractive indices of the two media and the direction of the incident ray, Snell's law can be used to compute the refracted ray's direction (Figure 3.6).

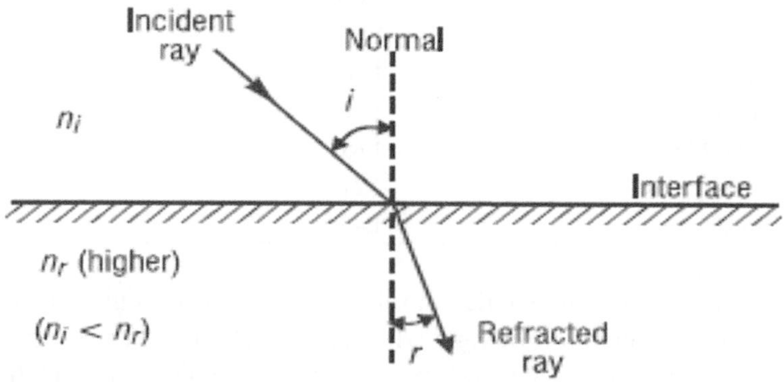

FIGURE 3.6 Principle of Snell's law.

$$\frac{\sin i}{\sin r} = \frac{n_r}{n_i}$$

where i is the incidence angle; r is the refraction angle; n_i is the index in the incident medium; and n_r is the index in the refracting medium.

Snell's law is written simply as:

$$n_i \sin i = n_r \sin r$$

3.1.7 CRITICAL ANGLES AND TOTAL INTERNAL REFLECTIONS

In Figure 3.7, four light beams from point O in the higher-index medium are shown, each incident on the interface at a different angle. Ray 1 makes contact with the object at a normal incidence of 90?, hence no bending takes place. The light in this path accelerates but remains in the same direction in the second medium. Ray 2 bends away from the normal at angle r after colliding with the surface at angle i. Ray 3 encounters the critical angle i^c, which is significant enough to cause the refracted ray to traverse along the interface (*N*) between the two media by 90?. Ray 4 obeys the rule of reflection; hence its angle of reflection equals its angle of incidence.

When engineering light propagation in fibers, we take advantage of the phenomenon of complete internal reflection by trapping the light in the fiber by repeated internal reflections along the fiber. The clarity and brightness of entirely internally reflected light beams are enhanced when compared to regular reflection from mirrors.

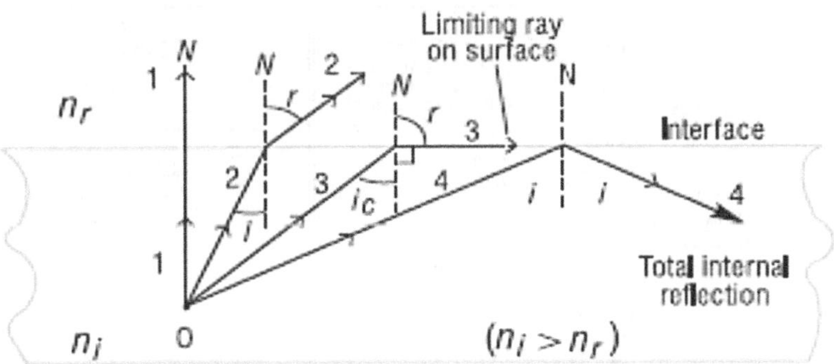

FIGURE 3.7 Total internal reflections.

Snell's law is used to calculate the critical angle of incidence for any two optical media when light is incident from the medium having a higher index. Using Snell's law and Ray 3 in Figure 3.7, we get:

$$n i \, sin \, i_c = n_r \, sin 90□$$

where; n_i is the index of the incident medium; i_c is the critical angle of incidence; n_r is the index of the medium with lower refractive index, and the angle of refraction at the critical angle is 90. The critical angle is therefore obtained since sin 90? = 1:

$$i_c \quad sin^{1}\left(\frac{n_r}{n_i}\right)$$

This phenomenon describes the basics of optical fiber operation.

3.1.8 *CURVED SURFACE'S REFLECTION*

When reflecting light at curved surfaces using spherical mirrors, the process appears to be similar to that in optical fibers. The laws of reflection hold because, as illustrated in Figure 3.8, for any point on the curved surface, a tangent may be drawn and a normal can be constructed to a point *P* on the surface of incident light. The laws of reflection are then applied at point *P*, as shown in Figure 3.8, with the incident and reflected rays forming identical angles (angle *B* equals angle *A*) with the normal to the surface at *P*. Surface tangents along the curved surface in Figure 3.8 are ordered (not random) portions of plane mirrors that act as a spherical surface when smoothly joined—as a spherical surface mirror, capable of forming different images.

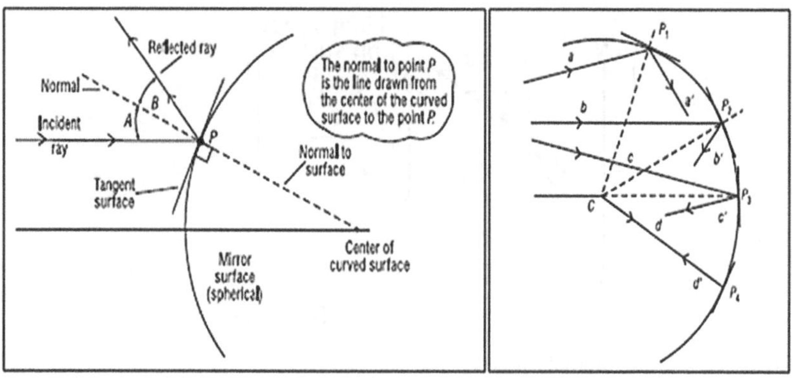

FIGURE 3.8 Reflections at curved surfaces: angle B equals angle A.

The reflected ray's direction can always be determined by applying the laws of reflection, since point *P* may be moved anywhere along the curved surface and a normal drawn there.

3.1.9 NUMERICAL APERTURE

It is a key lens-design characteristic that is directly tied to determining the amount of light captured by the lens. As the focal length and diameter of a design lens increase, so does the solid angle (cone) of usable light rays from object to image. The design of the objective lens (the lens adjacent to the specimen) for a microscope is an application of numerical aperture. Light-gathering capabilities are critical for microscopes.

Figure 3.9 displays the light-gathering capacity of a lens in relation to a point *O* on a glass-slide-covered specimen. Lens *L* is the microscope's objective lens that is focused on the specimen. The light-gathering capability of the lens, with air between the cover slide and the lens, is illustrated in terms of the half-angle (α_{air}) with air on the right side of the lens's symmetry axis. On the left side, the improved light-gathering ability of the lens is illustrated in terms of the greater half-angle (α_{oil}) with oil located between the cover slide and the lens. The oil is designed to have a refractive index (n_0) that is extremely close to that of the cover slide (n_g), resulting in little or no refraction at the glass–oil interface for limiting beam 2.

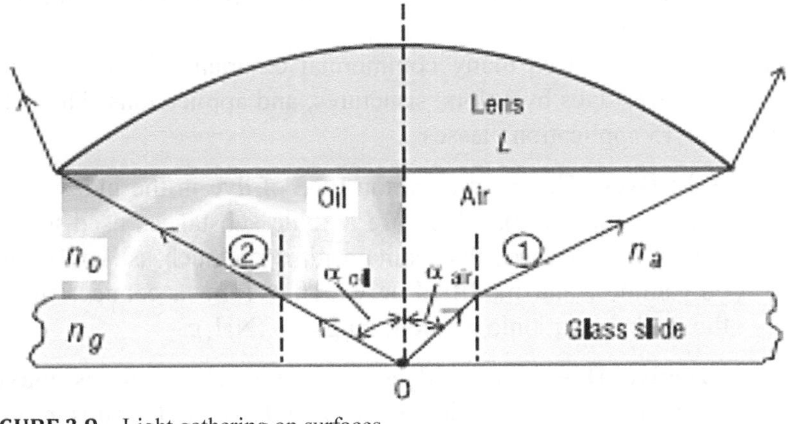

FIGURE 3.9 Light gathering on surfaces.

As a result, the half-angle (α_{oil}) is larger than the half-angle (α_{air}). As seen in Figure 3.9, ray 1 is refracted at the glass–air contact, limiting the received cone of rays by the lens to a smaller half-angle (α_{air}). A lens's numerical aperture is specified to show the difference in solid angles (cones) of accepted light, for example, by oil-immersion systems versus air-immersion setups.

The numerical aperture (N.A.) is defined as follows:

$$N.A. = n \sin a$$

where n is the index of refraction of the medium between the object and the lens, and the half-angle is determined by the limiting ray (α_{air} or α_{oil}). Raising the refractive index of the intervening medium boosts the light-gathering power of the microscope's objective lens accordingly. Because the entering rays to a fiber face are in the presence of air, the numerical aperture is:

$$N.A. = \sin a$$

3.1.10 DYES' CATEGORIZATION

All aromatic compounds absorb electromagnetic radiation, but only those that absorb in the visible region (350–700 nm) are colored. The dyes comprise chromophores, which are essentially double bonds in the structure, as well as auxochromes that are either electron-donating or electron-withdrawing and, when combined with chromophores, magnify the chromophore's color. The following are some of the most frequent dye types (Figure 3.10).

Because there are so many commercial colorants, the Color Index (C.I.) categorizes dyes by colors, structures, and applications. The C.I. is divided into 15 application classes:

1. **Acid Dyes:** The most common type of dye in the index is acid dyes (about 2,300), which are anionic substances used to color materials and yarns that contain nitrogen, such as wools, silks, polyamides, and modified acryl. They form a connection with those fibers' cationic ammonium group (NH_4).
2. **Reactive Dyes:** These dyes include reactive groups making covalent connections with the –OH, –NH, or –SH groups in fibers including cotton, silk, wool, and nylon. Typically, the reactive

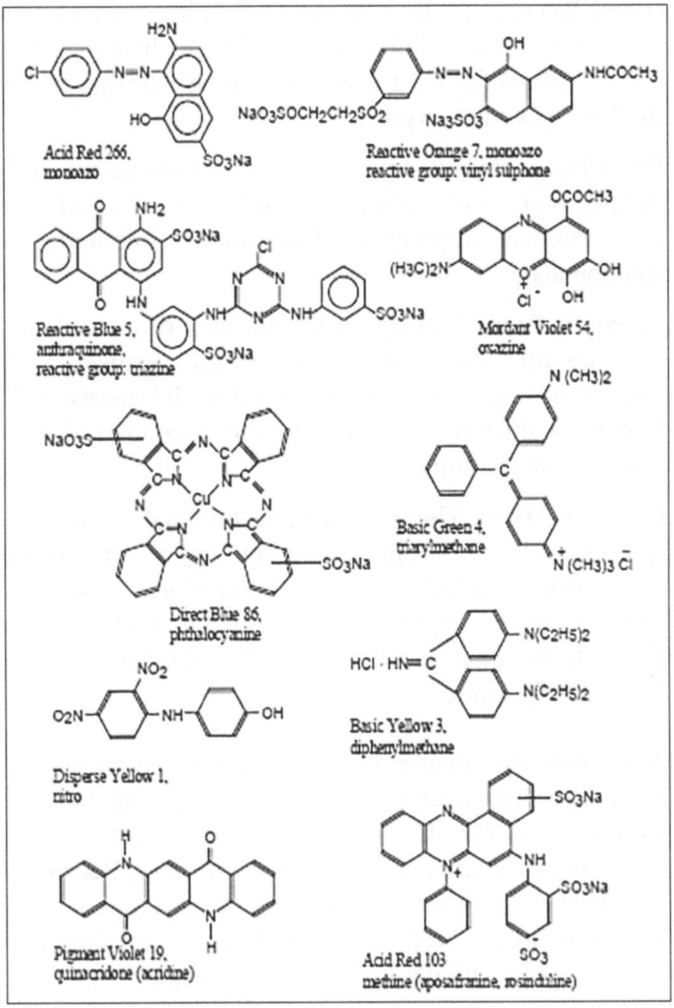

FIGURE 3.10 Most frequent dye types.

groups are heterocyclic aromatic rings replaced with fluoride or chloride. This is the second-largest group in the C.I.

3. **Metal Complex Dyes:** This category includes acidic and reactive dyes that have strong complexes comprising a single metal atom (cobalt, copper, chromium, or nickel) plus one or two dye molecules. Typically, metal-complex dyes are azo compounds.

4. **Direct Dyes:** These dyes are made up of big molecules that have a strong affinity for cellulose fibers. They are held together by weak van der Waals forces. These colors are generally azo compounds (R–N=N–R) with several azo bonds.
5. **Basic Dyes:** Because they are cationic compounds, this class of dyes is used to color fibers with acidic groups, such as modified polyacryl. The majority of them contain anthraquinone or azo compounds.
6. **Mordant Dyes:** Usually by the addition of a mordant, mordant dyes are attached to clothes. In layman's terms, a mordant is a binding chemical for the dye to the fabric. It is used to color wools, leathers, and cellulose fibers. Mordant is composed of chromium or dichromate complexes.
7. **Disperse Dyes:** These dyes have minimal value for coloring synthetic fabrics, as opposed to cottons, which may be colored with sulfur, reactive, azoic, and vat dyes. Disperse dyes are nonionic and dye synthetic fibers such as acetate fibers using a direct dyeing process. The only difference between fibers is the dyeing temperature.

   A nonionic dye, disperse dyes are moderately insoluble in water at ambient temperature and have limited solubility at higher temperatures. The dye is normally present in the dye bath as a fine aqueous solution with a dispersion agent. Because the water dissolves a small amount of dye, the hydrophobic fibers can absorb the dye from the solution. The dyes are organic nonionic chemicals of relatively low molecular weight.

   Simple and moderately insoluble *azo* and *anthraquinone* compounds were dispersed in water using a sodium salt as the first disperse dyes. A thin dispersion is essential for quick dyeing and also prevents large dye particles from adhering to the cloth.

   Because of the presence of polar substituents on the molecular structure of disperse dyes, a small amount of dye becomes present in a true aqueous solution. This aqueous solution may permeate hydrophobic synthetic fibers. Because of this phenomenon, the dye is significantly more soluble in the fiber than in water, allowing for deeper dye penetration. The following describes the solubility equilibrium:

$$Dye_{(s)} \rightleftharpoons Dye_{(aq)} \rightleftharpoons Dye_{(fiber)}$$

Fastness to washing and light of fibers or textiles colored with disperse dyes is moderate to good. The acrylic fibers can be dyed using disperse dyes, although the dyeing extent is restricted and the color build-up is poor. Because of the effect of high heat on the fabric, dyeing at temperatures above 100°C to fix this problem is not practicable. Acrylic fiber dyeing using disperse dyes is confined to faint colors due to a variety of causes.

Small *azo* or *nitro* compounds (yellow to red), *anthraquinones* (blue and green), or *metal-complex azo* compounds are commonly used in dyes. Some examples of typical disperse dye structures are given in Figure 3.11.

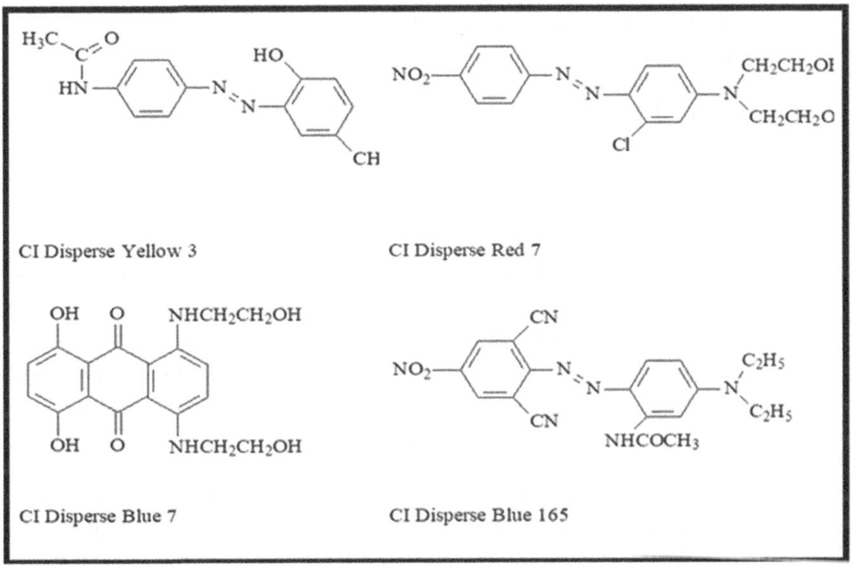

FIGURE 3.11 Typical dispersion dye structures.

8. **Pigment Dyes:** Organic pigments, often known as pigment dyes, make up a minor portion of the pigments used in dyeing. The dye consists of nonionic compounds or salts that are insoluble and

retain their structure during application. It is used by dispersing it in an aqueous solution, which, of course, requires a dispersing agent. The majority of pigment dyes are *azo* compounds (red to yellow) or metal complexes (green and blue).

9. **Vat Dyes:** This series of dyes includes water-insoluble dyes that are often employed on cellulose fibers. The dyeing procedures are based on the vat dyes' solubility in their reduced state (reduced with sodium dithionite), after which oxidation is permitted to return the dye to its original form once impregnated on the cloth. The majority of vat dyes are *indigoid* or *anthraquinone* compounds. One of the oldest instances includes *indigo* for vat dyeing.
10. **Ingrain Dyes and Anionic Dyes:** Ingrain dyes and anionic dyes, or *naphthol dyes*, are insoluble by-products of a reaction between a coupling component such as phenols, acetoacetylamides, or naphthols (*C.I. Azoic Coupling Components*) and a diazotized aromatic amine (*C.I. Azoic Diazo Components*). The reaction occurs on the fiber. Most of the naphthol dyes are regarded as *azo* compounds.
11. **Sulfur Dyes:** Sulfur dyeing includes reduction and oxidation, similar to vat dyeing. They are mostly used to color cellulose fibers.
12. **Solvent Dyes:** Solvent dyes, or lysochromes, are nonionic dyes used for coloring surfaces that may dissolve them, such as waxes, ink, fats, varnish, and plastics. They are not commonly employed in textile processing, although their use is growing. The majority of solvent dyes are *diazo* compounds that have undergone molecular restructuring. Solvent dyes such as *anthraquinone*, *triarylmethane*, and *phthalocyanine* are also used.
13. **Fluorescent Brighteners:** Fluorescent brighteners (or bluing additives) absorb ultraviolet light and emit weakly visible blue light, masking the yellowish color of natural fibers. They are not dyes in the traditional sense since they lack vibrant color. Several groups of fluorescent brighteners are distinguished based on their chemical structures: *stilbene derivatives*, *coumarin derivatives*, *pyrazolines*, *naphthalimides*, *1,2-ethene* derivatives, and aromatic or heterocyclic ring structures. *Triazinyl* units

and water-solubilizing groups are found in several fluorescent brighteners.

14. **Food Dyes:** In addition to the dye classes indicated above, the *Color Index* (C.I.) includes food dyes. These dyes are used primarily in the food industry and are rarely employed in textile processing activities.
15. **Natural Dyes:** These which include compounds such as *indigoid, anthraquinone, flavone, flavanol,* or *chroman* compounds, can be employed as mordant, vat, direct, acid, or solvent dyes. These dyes are seldom utilized in textile processing activities.

3.1.11 MEASUREMENT OF COLOR

Color measurement is critical in textile dyeing, particularly for sample matching, identifying color discrepancies, and preparing dye baths. Fundamentals of color measurement, often known as *colorimetry*, are necessary for dyeing technology.

Colorimetry is the quantitative description of colors using physical measures. Two samples with the same numerical color specification under the same viewing circumstances will always have the same color perception under those conditions. Color differences should be assessed, and allowable differences should be maintained at a minimum. Colorimetry is used in many businesses producing-colored clothes or products, making it an important method in textile dyeing.

3.1.12 INFLUENCING FACTORS IN COLOR PERCEPTION

It is difficult to answer the question of color and how we see it. Color perception is a complex hypothesis that is tied to light absorption in the eye and processing in the brain. In textiles, the material selectively absorbs some of the illuminating light while transmitting or reflecting the remainder to the observer's sight (Asada and Yuuki, 1994; Bayle and Mateo, 1996; Blyler, 1999; Ioannides et al., 1996; Steiger et al., 1998; Zoido, 1998). The light that enters the eye is what gives the observer (human) the sensation of colors.

Light perception might differ depending on three factors:

- The wavelength distribution or the light source's color for lighting the item;
- The degree of reflections or transmissions of each incident light's wavelength by the object; and
- The observer's visual reaction to the light's wavelength arriving from the object.

The numerical definition of a color necessitates the numerical descriptions of each of the contributing components listed above.

3.1.13 LIGHT SOURCES WITH ILLUMINANTS

- White light, as opposed to specialized colored light like red light, contains all visible-light wavelengths in about similar amounts, making it the greatest supplier of color rendering qualities.
- Line spectra with discontinuities from a source that only produces light's discrete wavelengths, such as a low-pressure mercury lamp.
- Emission spectra from sources emitting at all wavelengths, such as an incandescent tungsten lamp or the sun.
- Colored lights with limited color-rendering capacity, which are typically white lights.
- Both line and continuous emissions, such as fluorescent tube light, frequently provide high-quality white lights.
- A light source's spectral power distribution (SPD) provides the necessary numerical description. SPD expresses the radiated power as a function of wavelength ($W\ m^2\ nm^{-1}$).
- Daylight, which has an uneven, incessant SPD, is a white light consisting of all visible wavelengths. An incandescent light source (tungsten lamp) is more yellow than daylight and hence has a more consistent SPD. Fluorescent light produces an erratic SPD.
- The color temperature, as specified by the absolute temperature scale in Kelvin (0°C = 273.15K), is commonly used to categorize light sources. The greater the total powers of the emitted radiation and the shorter the wavelength of the maximum emission, the higher the temperature of the emitting source.
- SPDs were recommended by the *Commission Internationale de l'Éclairage* (*CIE*) for light sources used in colorimetry, known as

CIE standard illuminants. A *CIE Illuminant D65*, for example, has a set SPD for a number of illuminants comparable to daylight. *D* symbolizes daylight, and the number signifies the color temperature in hundreds of degrees Celsius (65 = 6500 K).

3.1.14 SPECTROPHOTOMETRY REFLECTION

Spectrophotometry describes the numerical transmission or reflection of light by an object, showing the amount of incident light that an item transmits or reflects as a function of wavelength. Most reflection spectrophotometers assess the sample's light reflection at each wavelength compared to a white standard, such as a plate coated with $BaSO_4$ or MgO. The standards diffuse nearly 100% of the reflection between 380 and 750 nm. Some spectrophotometers have pulsed or continuous light sources, varied angles for incoming and reflected light beams, double or single beam optics, different dispersion systems into component wavelengths, and variable wavelength ranges and measuring frequency. Figure 3.12 shows some familiar spectrophotometer geometries.

The following are the three most frequent instrument geometries:

- 45? angle illumination of the sample's surface and measurement of light reflected at nearly right angles;
- Opposite of the above; and
- The sample is diffusely illuminated from all angles using an integrating sphere, and reflected light that is almost perpendicular to the sample is detected.

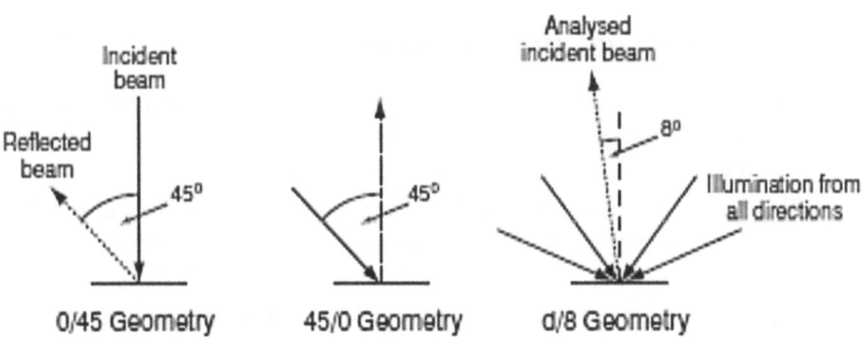

FIGURE 3.12 Familiar spectrophotometer geometries.

Certain spectrophotometers can constantly scan the whole visible spectrum, while others only measure at certain wavelengths. Because of differences in equipment technology, findings for similar samples are not identical.

3.1.15 MEASUREMENTS OF REFLECTANCE IN TEXTILES

Texture and color variations in textile fabrics are caused by manufacturing processes. As a result of these circumstances, during manufacturing, samples are obtained from various areas of a batch, and the reflection spectrum of each sample is measured multiple times. This study might yield the average reflectance spectrum. The sample should be opaque for reflectance measurements, which means that no light should pass through it. This may cause light to bounce back from the sample holder, which is undesirable; thus, the cloth may be layered to prevent this.

Almost all fabrics have a distinct texture that influences the dispersion of incident light, changing the appearance from various perspectives. To reduce this impact, the fabric should be rotated in orientation by 90? between analyses, and an average should be obtained, limiting the effects of texture on direction.

Conditions should be kept under control to ensure repeatability of outcomes. Temperature and humidity on the sample might cause variations in reflectance. Light sources are also significant in reflectance studies; for example, when there is a small amount of ultraviolet light (wavelengths below 400 nm), fluorescent dyes can be excited.

3.1.16 POLYMETHYL METHACRYLATE

The monomer of poly(methyl methacrylate) is methyl acrylate, and the polymer is generated by simple alkene addition; the reaction was adapted from *Clayden's Organic Chemistry* (Figure 3.13):

"PMMA stands for poly(methyl methacrylate)." It is a robust and lightweight substance (Blyler, 1999; Ishigure et al., 2002; Morisawa and Muto, 2012; Zoido, 1998) with a density half that of glass (glass's density ranges between 2.4 and 2.8 g/cm^3). When compared to glass and polystyrene, it also has high impact strength.

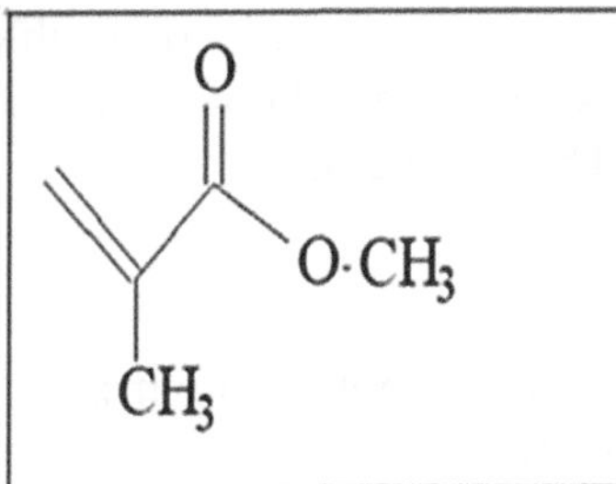

Polymethyl methacrylate	
Chemical Formula	$C_5H_8O_2$
Molar mass	100.12g.mol^{-1}
Density	0.94 g/cm^3
Melting point	-48°C, 225K
Boiling point	101°C, 374K

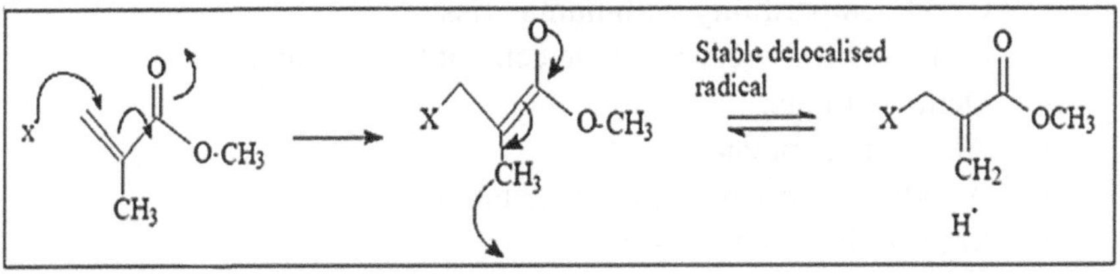

FIGURE 3.13 Radical formation of alkene addition.

The polymer can be easily formed after the radical formation as shown in Figure 3.14.

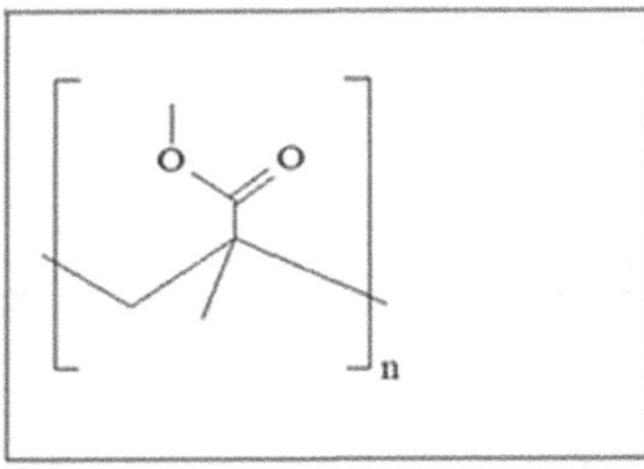

Polymethyl methacrylate	
Molecular Formula	$(C_5H_8O_2)_n$
Density	1.18 g/cm^3
Melting point	160°C, 433K
Boiling point	200°C, 473K
Refractive index	1.4914 at 587.6nm

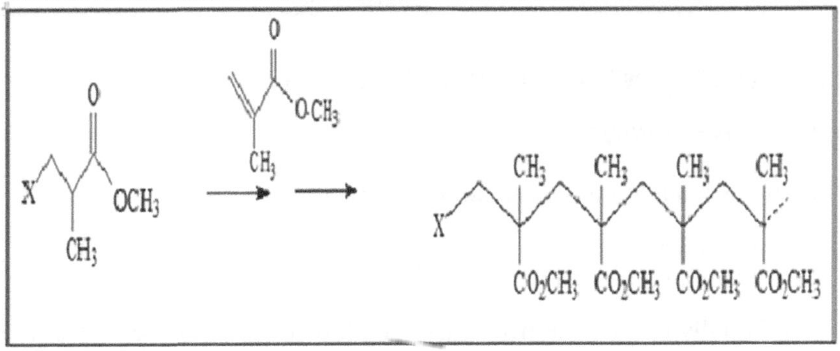

FIGURE 3.14 Formation of poly(methyl methacrylate).

Due to its refractive index, PMMA transmits up to 92% of visible light and reflects roughly 4%. The polymer is very hydrophobic since the maximum water absorption ratio is 0.3 to 0.4% by weight.

Among the applications are:

- Transparent glass replacement;
- Daylight redirection (in skylights, to spread light in a room);
- Medical implants and medical technology;
- Good compatibility with human tissue;
- Orthopedic surgery (as bone cement to fix implants);
- Dental fillings;
- Aesthetic applications;
- Acrylic colors (hydrophobic nature);
- Signs (advertising business);
- Plastic optical fibers;
- High flexibility and low cost are significant benefits for short-distance communication.

3.1.17 FIBER COMPOSITION

The fiber composition is as shown in Table 3.2.

TABLE 3.2 Fiber Composition

Properties of Fiber	Fiber Explanation
Flexi	1.5 mm
Composition of core	PMMA
Composition of cladding	Poly(tetrafluoro-ethylene)

3.2 EXPERIMENTAL SET-UP AND ANALYSIS

3.2.1 CORE DYEING

The fiber is made up of two major components:

- core; and
- cladding.

By initially cutting the necessary length of fiber, the core of the fiber had to be manually peeled (60 cm). Following that, tiny incisions were performed to liberate the core from the cladding on either end, releasing any vacuum produced in the fiber while cutting. Keeping the cladding in place on one end while tugging on the core from the other, the core was then removed. The apparatus setup with dye classification for fiber dyeing is shown in Figure 3.15.

A watch glass was used to weigh 5 g of Terasil Red GFF, which was then placed in a dye bath (30 cm × 25 cm, stainless steel) holding exactly 1 liter of boiling water. The temperature was measured after the addition of dye. Following that, the fibers were placed in the dye bath for their corresponding dye durations (10, 20, 30, 40, 60, 90, 120, 150, 180 minutes). To avoid changes in dye concentration, the water level in the boiling dye bath was maintained constant. After removing the fibers from the dye solution, they were thoroughly washed with water.

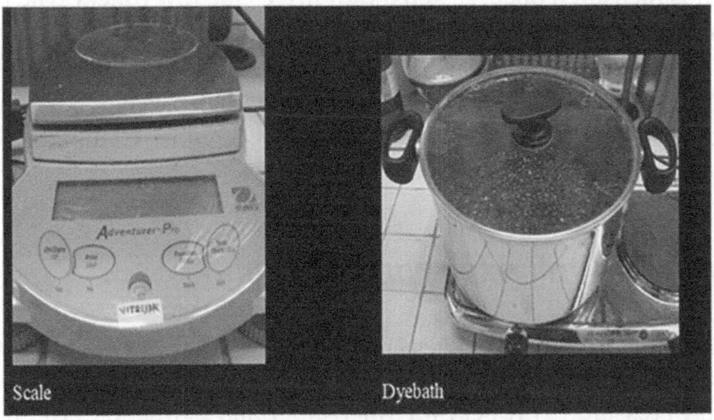

Dye Classification	
Trade name	Huntsman Terasil Red GFF
Dye concentration	5g/L
Temperature	98°C
Time	10, 20, 30, 40, 60, 90, 120, 150, 180 minutes

FIGURE 3.15 The apparatus setup for fiber dyeing.

3.2.2 *TEMPERATURE EFFECTS ON CORE AND PARENTAL FIBER*

For this study, 60 cm samples were produced. As detailed in the preceding section, a collection of core samples was prepared (i.e., Core PMMA and Parental Flexi). The parental fiber was cut to size but not altered in any manner. Six samples were immersed in a boiling water bath (with and without cladding), and a single fiber of each type was retrieved at 10, 20, 30, 40, 60, and 90 minutes at a temperature of 98–100°C.

3.2.3 FIBER SIDE EMISSION

3.2.3.1 PRINCIPLE

The device is designed to guide a laterally emitting optical fiber or a textile structure including optical fibers and to measure the light output emitted by these structures. The optical fiber or textile structure is positioned between the feed rollers, which direct them to the measuring tunnel, where the light output is measured once. A step motor drives the tow rollers. The thread is illuminated after being guided through the apparatus.

The actual measurement is carried out with the aid of a light sensor, which reads the light output in predetermined step lengths. A step motor powers the tow rollers and processes step lengths. Computer software written in MATLAB controls the device.

3.2.4 THE INSTRUMENT'S DESCRIPTION

According to Figure 3.16, the device comprises a base plate (1). This plate is attached to the margins of four columns (2), which include fixed-free (4) and pressure (5) rollers for fiber input and extraction. The poles are brackets (3), which are attached to a pressure roller that functions as a shaft lever (7). It enables roller manipulation. The real pressure roller is mounted on the push lever (6), the intensity of which is controlled by two tension springs (11). A fixed drive belt (12) driven by a step motor (9) runs alongside the bottom roller or driving roller (8). The base plate supports the center measuring tunnel (10). It is outfitted with two covers to prevent measurement bias. Furthermore, the measuring tunnel has two holes for mounting a spectrometer and a light sensor.

3.2.5 PLASTIC OPTICAL FIBER CROSS-SECTION ANALYSIS

After dyeing the samples, cross-sectional photographs were used to assess dye diffusion through the fiber (Arrue et al., 1998; Blyler, 1999; Daum et

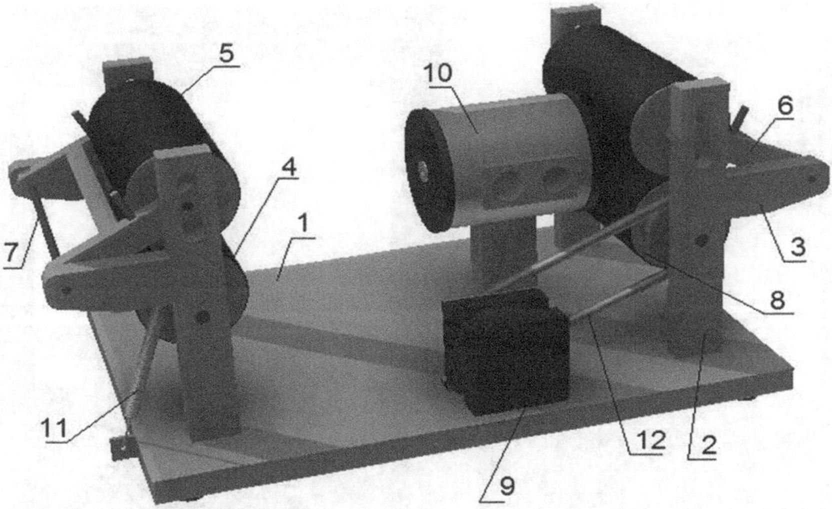

FIGURE 3.16 Schematic diagram of Prototype 1 for measurement of side emission.

al., 1994; Yabre, 2000). To ensure that the fiber remained straight, it was first cut to the exact size of the wax mold. The fiber was then wrapped with cotton roving, resulting in parallel cotton fibers. After wrapping the PMMA core, the cotton roving needed to be twisted slightly so that it completely covered the PMMA sample.

The sample was then placed in the wax mold, and the mold's sides were sealed with scotch tape. The paraffin wax was then heated to melting point and allowed to cool before being transferred to a small beaker (50 mL) to facilitate easy pouring of the warm wax into the mold (Koike and Ishigure, 2006). After filling the mold to the top with wax, it was kept overnight at temperatures below –5°C.

The wax was removed from the mold after it had been frozen overnight. Using a microtome tool, the sample—now encapsulated in wax—was sliced into 70 μm cross sections. Samples were sliced slowly, and the cross section was cleaned of any wax before being gently transported and placed onto a glass slide containing a thin layer of glycerin for examination (Figure 3.17).

The samples were then analyzed using NIS-Elements and a Nikon microscope at 5.00× magnification, with the specific calibration values listed in Table 3.3.

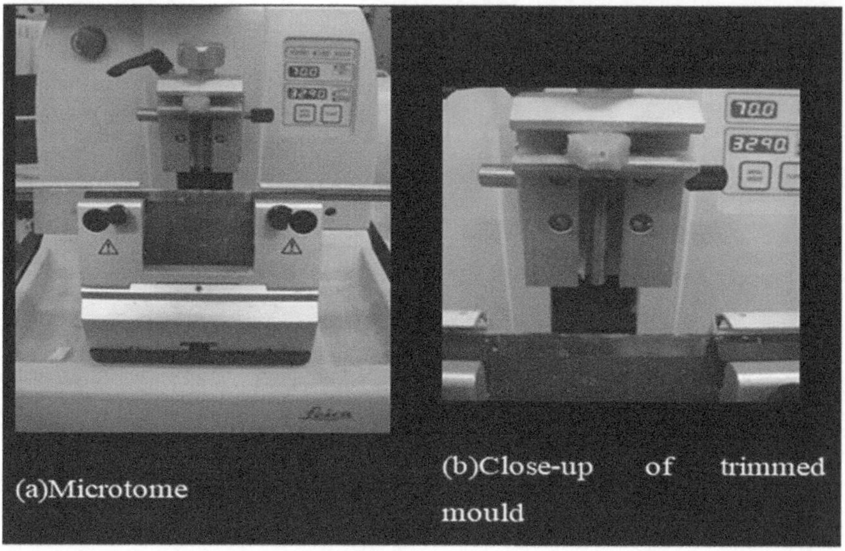

FIGURE 3.17 Fiber cross-section analysis.

TABLE 3.3 Calibration Settings for Light Microscope

Instrument	DS-5M-US
Magnification	5.00×
Format	640 × 480
Gain	1.70×
Exposures	ME 10 ms
White balance	1.02, 1.00, 3
Noise reduction	On
Saturation	0.19
Offset	0.00
Hue	0.19
Sharpness	Smoothest
Contrast	Dynamic

3.2.6 SURFACE TREATMENT WITH ETHYL ACETATE

To generate a surface alteration to the parent fiber, ethyl acetate was used as an etching solution. In a sealed container, 90% concentration was used to treat 1-meter samples of fibers for 1 hour and 30 minutes. It is vital to remember that the solvent is highly flammable and that precautions should be taken. Following solvent treatment, the fibers were thoroughly washed with water and allowed to dry at room temperature. Table 3.4 illustrates the solvent and sample details.

TABLE 3.4 Solvent and Sample Details

	Solvent
Ethyl acetate	900 g/1000 mL
Molecular formula	$C_4H_8O_2$
Molecular weight	88.11 g/mol
	Sample
Flexi	PMMA core and poly(tetrafluoroethylene) cladding

3.2.7 DYEING OF SURFACE-MODIFIED OPTICAL FIBERS

Dye baths were prepared at concentrations of 10 g/L and 15 g/L. The bath was allowed to reach boiling temperature (98°C), and then the treated fibers were placed into their respective dye baths. After 20 minutes of dyeing, the fibers were removed and rinsed with water. Table 3.5 illustrates the dye classification for surface-modified optical fibers.

TABLE 3.5 Dye Classification for Surface-Modified Optical Fibers

	Dye Classification
Trade name	Huntsman Terasil Red GFF
Dye concentration	5 g/L, 10 g/L
Time	20 minutes
Temperature	98°C

3.2.8 DYED FIBER CROSS-SECTION ANALYSIS

Initially, using a 5 g/L concentration, the dyeing of fiber cores was performed, followed by examination of fibers cut into 5 cm lengths under a light microscope at 5× magnification. Figure 3.18 illustrates that even after 3 hours of dyeing (i.e., after 240 minutes), light could still pass through the material. After 60 minutes, the fiber became exceedingly brittle and prone to breaking, according to the findings of this study. Additionally, the fiber stiffened, making dyeing impractical, as the suppleness of the fiber is essential for incorporation into fabrics.

This approach was also questionable due to the impossibility of using the same calibration parameters to assess dye penetration or fiber diameter. Figure 3.18 also indicates that a better method of cutting the fibers is required, as the surfaces must be polished more smoothly to obtain clearer microscopic images.

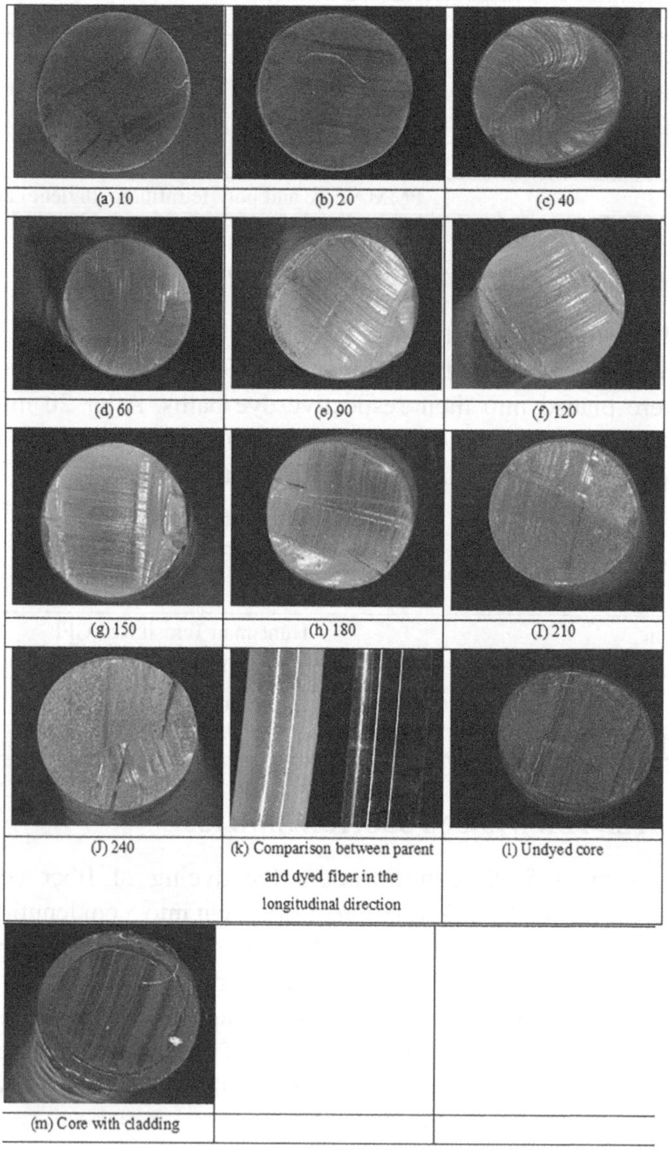

(Continued)

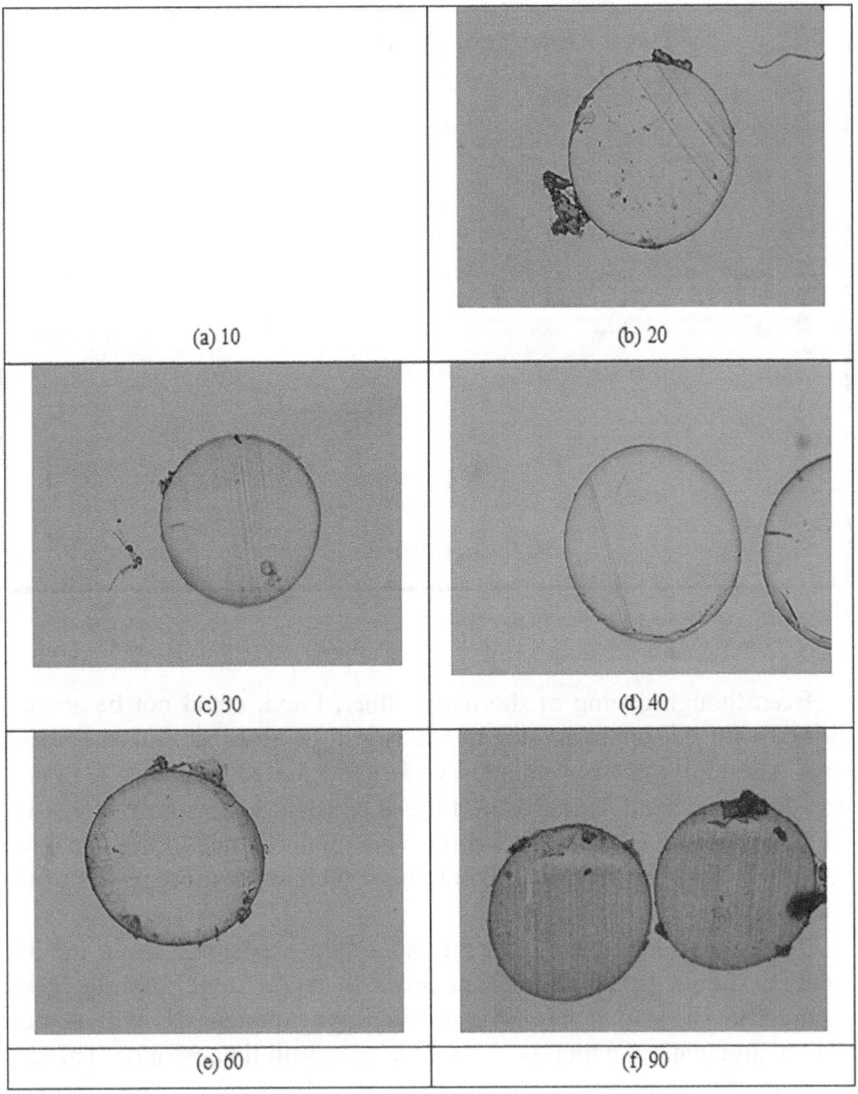

FIGURE 3.18 Dyed fiber cross-section analysis.

3.2.9 CORE DYEING

Subsequent to dyeing of the PMMA core and its analysis, the results obtained were as in Figure 3.19.

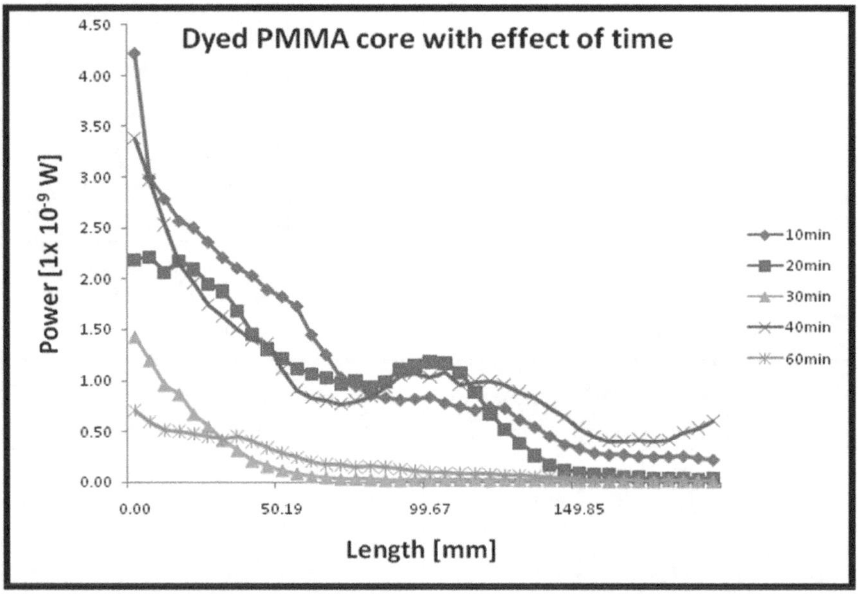

FIGURE 3.19 Effects of time in dye bath.

Even though dyeing of the parent fiber, Flexi, could not be accomplished with this approach, the dye may still be absorbed into the fiber's core. Although the fiber samples were dyed for 120 minutes at first, it was discovered that, as exposure time at constant temperature increased, the fiber became exceedingly brittle. This might be due to the dye's full permeation into the fiber and alteration of the chemical properties of the PMMA core.

It was later determined that the maximum exposure time in the dye solution should be 60 minutes in order to retain some fiber integrity. Figure 3.19 shows that side emission declines exponentially with respect to light attenuation, which is observable across all time periods. The side emission decreases progressively as the dye duration increases from 10 to 30 minutes, but it unexpectedly increases at 40 minutes. This might be because the dye is generating a suitable cladding for the PMMA core, allowing light to remain within the fiber for extended attenuation periods. For the next 60 minutes, the emission decreases once again. The effects of temperature on the parent and core of Flexi over time are shown in Figure 3.20.

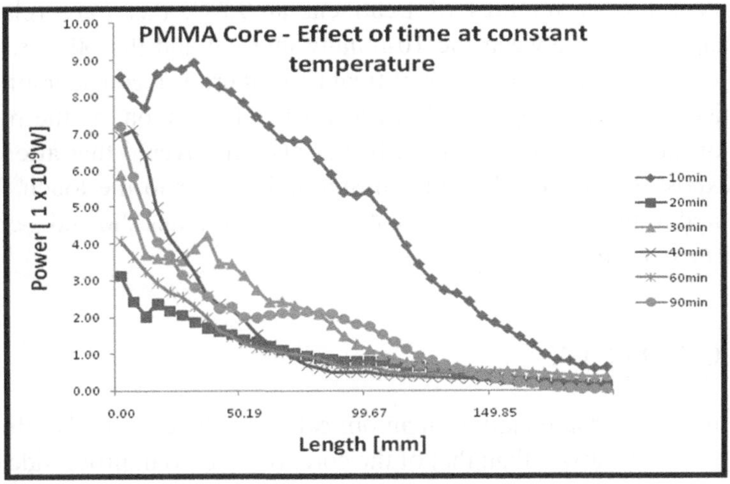

FIGURE 3.20 Effects of time at constant temperature for PMMA core.

After the shortest time duration of 10 minutes, the fiber exhibits the best attenuation and side emission. As time progresses, the light output gradually decreases until the 40-minute interval is reached. Similar to the dyed core, this phenomenon might be due to water penetration, causing part of the cross section to form a radial cladding, which enhances side emission and light attenuation. Power emission and attenuation decrease in proportion to time but increase again around 90 minutes. Temperature effects on Flexi over a period of time is given in Figure 3.21.

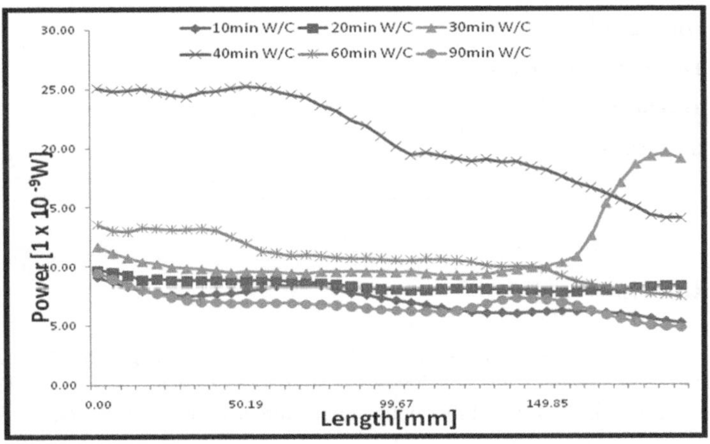

FIGURE 3.21 Temperature effects on Flexi at different time intervals.

Except for 40 minutes of treatment at 98°C, there are relatively minimal changes between the 10-minute interval and the other samples (Figure 3.21). After 90 minutes of treatment, there is a considerable drop in emission, which may be related to fiber deterioration, as the melting point of the fiber is about 160°C. It was also discovered that after some heat exposure, the fiber became substantially softer to the touch. When compared to the other samples, treatment at 40 minutes showed a considerable increase in softness.

3.3 CONCLUSION

To achieve better attenuation in an optical fiber, the cladding's refractive index should be lower than that of the core. In order to improve side emission, part of the internal reflection caused by the refractive index difference must be lost. As a result of the influence of constant temperature over time, there may be changes in the refractive index of either the cladding or the core.

KEYWORDS

- **dyeing**
- **fiber integrity**
- **flexi**
- **optical fibers**
- **radical formation**
- **refractive index**
- **side emission**

REFERENCES

Arrue, J., Zubia, J., Fuster, G., & Kalymnios, D. (1998). Light power behavior when bending plastic optical fibers. *IEE Proceedings-Optoelectronics*, *145*(6), 313–318.

Asada, K., & Yuuki, H. (1994). Fiber optic temperature sensor. *3rd POF*, *94*, 49–51.

Bayle, J. J., & Mateo, J. (1996, October). Plastic optical fiber sensor of refractive index, based on evanescent field. In *Proceedings of the Fifth International Conference on Plastic Optical Fibers and Applications—POF'96* (pp. 220–227).

Blyler, L. (1999). Material science and technology for POF. In *Proc. Eighth Conference on Plastic Optical Fibers and Applications–POF'99.*

Daum, W., Hoffman, A., Strecker, U., Brockmeyer, A., & Weinert, A. (1994, October). Influence of chemicals on the durability of polymer optical fibers. In *Proceedings of the 3rd International Conference on Plastic Optical fibers, Yokohama* (pp. 111–114).

Ioannides, N., Kalymnios, D., & Rogers, I. W. (1996, October). An optimized plastic optical fiber (POF) displacement sensor. In *5th International Conference on Plastic Optical Fibers and Applications* (pp. 251–255).

Ishigure, T., Tanaka, S., Kobayashi, E., & Koike, Y. (2002). Accurate refractive index profiling in a graded-index plastic optical fiber exceeding gigabit transmission rates. *Journal of Lightwave Technology, 20*(8), 1449–1456.

Koike, Y., & Ishigure, T. (2006). High-bandwidth plastic optical fiber for fiber to the display. *Journal of Lightwave Technology, 24*(12), 4541–4553.

Morisawa, M., & Muto, S. (2012). Plastic optical fiber sensing of fuel leakage in soil. *Journal of Sensors, 2012.*

Steiger, J. U., Nickerson, P. W., Hermle, M., Thiel, G., & Heim, M. H. (1998). Interferon-γ receptor signaling is not required in the effector phase of the alloimmune response. *Transplantation, 65*(12), 1649–1652.

Yabre, G. (2000). Theoretical investigation on the dispersion of graded-index polymer optical fibers. *Journal of Lightwave Technology, 18*(6), 869.

Zoido, H. G. G. J. (1998). Mechanical properties of polycarbonate optical fibers. *Fiber & Integrated Optics, 17*(3), 231–242.

Zubia, J., & Arrue, J. (2001). Plastic optical fibers: An introduction to their technological processes and applications. *Optical Fiber Technology, 7*(2), 101–140.

CHAPTER 4

Modeling of 2D and 3D Woven Structures and Properties

HAFSA JAMSHAID*

Faculty of Textile Engineering, National Textile University, Faisalabad, Punjab, Pakistan

*Corresponding author

ABSTRACT

The selection of fabric characteristics has a significant impact on fabric attributes. The structure is influenced by the fabric properties. The behavior and connection of fabric parameters (FPs) is a prerequisite to finding the best solution for fabric engineering challenges. Many aspects of the textile are fundamentally determined by geometrical connections. The fabrics' geometrical models provide simplified equations to aid computations and specific constants used in textile engineering, structural analysis, and mechanical evaluation. These FPs serve as a useful tool for a creative fabric designer when producing materials for a variety of uses. Because of the theoretical link between the FPs, the designer can experiment with different fibers, yarn tex, threads per cm, and weave to modify texture and FPs.

4.1 INTRODUCTION

Finding the optimum solution for fabric engineering difficulties requires understanding the behavior and interaction of fabric characteristics. Geometrical linkages are essentially responsible for many elements of textile design. The fabrics' geometrical models provide simplified equations to assist in computations, as well as specific constants used in textile engineering, structural problems, and mechanical properties. The subsections describe plain weave and non-plain weave fabrics.

Advances in Industrial Manufacturing and Materials. Integrating a New Era of Computational Modeling and Machine Learning. Debesh Mishra and Swagatika Mishra (Eds.)

DOI: 10.1201/9781779640475-4

4.1.1 PLAIN WEAVE FABRICS

Peirce (1937) completed the first research on fabric structural design, which is the Peirce model of plain weave fabrics with yarns having a circular cross section. In addition, he presented a fabric model with an elliptic yarn cross section. Kemp (1958) introduced a racetrack concept. Hearle and Amirbayat (1988) suggested lenticular geometry for computations in fabric mechanics using the energy technique. Similarly, Grosberg and Kedia (1966), Backer (1952), and Postle et al. (1988) conducted several studies on fabric mechanical characteristics based on fabric structural models. Lindberg et al. (1961) conducted substantial research on the mechanical behavior of fabrics as it relates to tailorability. Kawabata et al. (1982) then developed an advanced technique for measuring mechanical characteristics of fabric, known as the KES-FB system. The CSIRO in Australia created the FAST fabric mechanical measuring system (Ly et al., 1991).

For examining fabric geometrical features, new objective measuring technologies (Hu, 2004), such as the virtual image display system (VIDS) and Fabric Surface-Analysis System, have recently been developed. On the other hand, various CAD systems (http://www.i-designer.co.kr) are now available for fabric design, including weave structures, colors, and patterns. There is also pattern-design CAD (http://www.texclub.com; Kuwabara, 1986), which includes a visual wear system (VWS) for clothing designers. However, there is currently no fabric structural design approach for determining fabric density based on fiber materials, yarn linear density, and weave patterns. As a result of design plan analyses for nylon and polyester textiles, a database system that can readily determine warp and weft densities based on various yarn counts, weave structures, and materials is required (Kim, 2007; Kim and Kim, 2010).

Soft computing may be used to anticipate FPs and their effects on fabric qualities, which enables decision-making based on specific buyer requirements. To solve these equations and find correlations between essential fabric properties—such as thread spacing and crimp, fabric cover and crimp, and warp and weft cover—a simpler approach is utilized. Such correlations aid in determining the direction in which FPs should be adjusted. For instance, the characteristics of a fabric needing a specific crimp in the warp and weft may be determined even while the cloth is jammed. A nonjammed cloth with a particular warp and weft quantity may also be desired. Fabrics having the highest crimp in either the warp or weft direction can be calculated. This effort attempts to optimize the engineering features of a plain weave cloth based on requirements.

4.1.2 NON-PLAIN WEAVE FABRICS

Section 1.1 explored all of the possible structures for plain weaves exclusively. The limiting structure was jammed structures and textiles that have maximum crimp in one direction and zero crossing thread crimp (or zero crimp in one direction). To envision beyond this condition—for example, to increase mass and fabric cover beyond what is achievable in plain weave—one must consider non-plain weave jammed structures, which extend the range of plain weave jammed structures beyond their limit. The weavability limit is defined as the maximum number of ends and picks per unit length that may be woven with a particular yarn and weave. Most research in this area has relied on empirical correlations (Dickson, 1954; Seyam and Harrison, 2002). The plain weave geometrical model is highly useful in forecasting this limit under jammed conditions for a given warp and weft yarn diameter (tex).

Fiber-reinforced composite materials offer several advantages and are therefore widely employed in a diverse range of products (Mallick, 2007). Because the majority of them are two-dimensional (2D) laminated composites, interlaminar delamination is the weakest failure mechanism in such materials. Several strategies for improving composite laminate strength have been proposed, with three-dimensional (3D) textile technology being the most successful (Chou, 1992; Ko, 1989). Several varieties of 3D textiles have been developed, each with its own unique characteristics. The orthogonal fabric composite, for example, presents a fiber architecture that aims to maintain in-plane performance while boosting interlaminar toughness by inserting only a modest quantity of through-thickness (*z*) reinforcements (Guénon, 1987; Guénon et al., 1989; Tanzawa et al., 1997, 1999).

The maximum weavability limit is anticipated using soft computing and by extending Peirce's geometrical model for non-plain weaves with circular and racetrack cross sections. This knowledge assists weavers in avoiding attempts to weave impossible structures, saving both time and money. It also aids in anticipating weaving difficulties and taking necessary warp preparation steps. Soft computing allows fabric designers to broaden the spectrum of fabric parameter (FP) selection for innovation. The yarn diameters and weave determine maximum weavability. Several factors influence yarn diameter, including fiber type, blend composition, and the spinning process. As a result, accurately estimating yarn diameter is crucial for forecasting weavability. Similarly, non-plain weaves must be characterized by appropriate numerical descriptors to support soft computing. This is accomplished through the use of the weaving factor concept.

4.2 METHODOLOGY

4.2.1 FOR PLAIN WEAVE FABRICS

4.2.1.1 APPROACH TO DEVELOP SOFT COMPUTING ALGORITHM

It goes without saying that Peirce's pioneering work on geometrical models provides the foundation for fabric engineering. Using contemporary soft-computing technologies, several basic fabric dimensional characteristics may be calculated from the geometrical model. FPs may be analyzed for any connections between warp and weft cover. Last but not least, the connection between fabric mass and fabric cover may be assessed. This is the power of soft computing that cannot be matched by any other method, such as graphical or nomogram methods.

The basic equations generated from the geometrical model are difficult to understand. In reality, solving these equations is challenging in many situations. Many researchers have attempted to solve the equations and have discovered some simpler correlations. However, the graphical strategy has been shown to be the most effective so far. As a result, it is decided to create a simple algorithm based on the Peirce geometrical model in order to ascertain the linkage between various fabric characteristics by the use of soft computing.

4.2.1.2 PROGRAMMING

MATLAB 7 is used to create computer programs for establishing correlations between various fabric properties derived from the Peirce geometrical model.

Peirce's geometrical relations can be expressed as follows:

$$\frac{p_2}{D} = (K - \theta_1)\cos\theta_1 + \sin\theta_1 \quad (4.1)$$

$$\frac{h_1}{D} = (K - \theta_1)\sin\theta_1 + (1 - \cos\theta_1) \quad (4.2)$$

where; p is the thread spacing; h is the crimp height; d is the yarn diameter; $D = d_1 + d_2$; l is the modular length; $K = l_1/D$; θ is the weave angle; and suffix 1 and suffix 2 indicate warp and weft yarn, respectively.

By replacing suffix 1 with suffix 2 and vice versa, two equivalent equations will be derived for the weft direction. The relations between FPs are graphically determined for various structures such as jammed structures, non-jammed structures, and fabric with straight cross threads. Figure 4.1 depicts the flowcharts for this method.

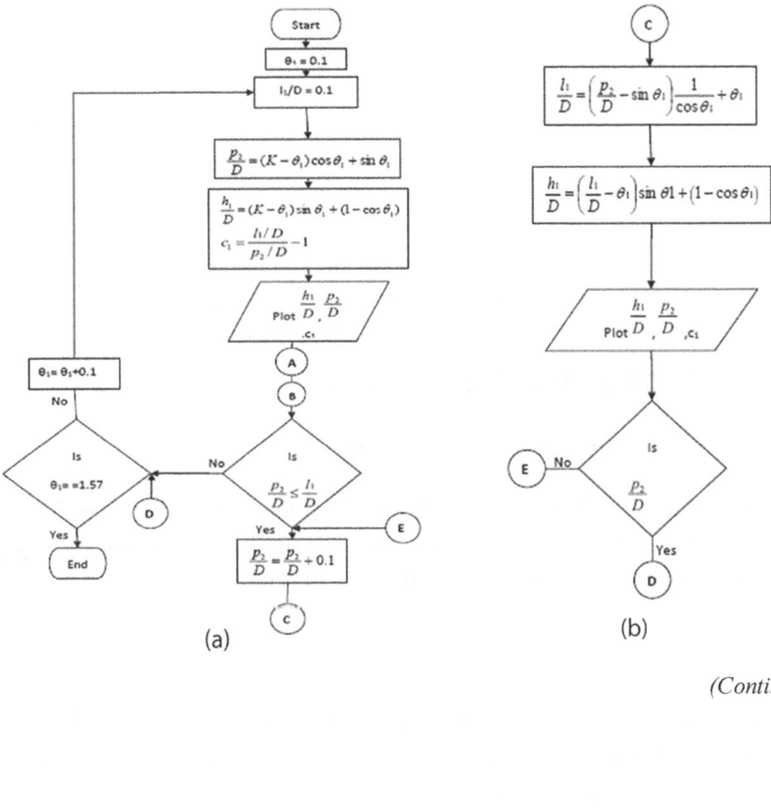

(Continued)

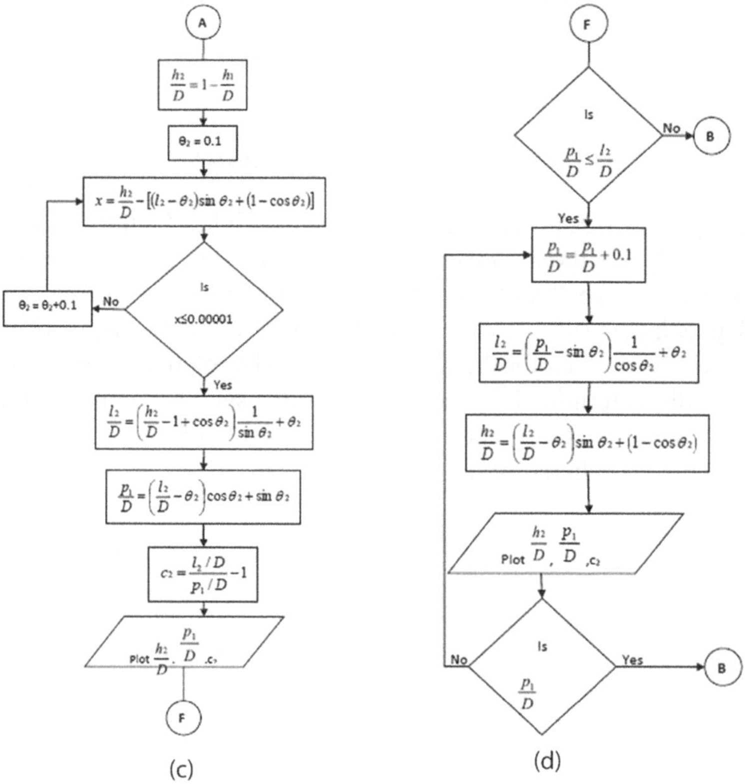

FIGURE 4.1 Flowcharts (a)–(d) to solve Peirce's basic geometrical equations.

4.2.2 *FOR NON-PLAIN WEAVE FABRICS*

4.2.2.1 YARN-DIAMETER ESTIMATION

To estimate weavability in general, two key geometrical parameters are required: yarn diameter and weave factor. For a general case, the yarn diameter in terms of linear density in tex is given as:

$$d = \frac{1}{280.2}\sqrt{\frac{T}{\phi\rho_f}} \tag{4.3}$$

where; d is the yarn diameter in cm; T is the yarn's linear density in tex (i.e., g/km); ρ_f is the fiber density in g/cm^3; ρ_y is the yarn density in g/cm^3; and ϕ is the yarn's packing coefficient.

This yarn-diameter equation is applicable to every yarn and fiber type. Fiber characteristics such as tex, length, fiber crimp, and cross-sectional shape influence the packing factor.

The following equation gives the average fiber density of mixed yarns:

$$\frac{1}{\overline{\rho}} = \sum_{1}^{n} \frac{p_i}{p_{fi}} \tag{4.4}$$

where; $\overline{\rho}$ is the average fiber density; p_i is the ith component's weight fraction; p_{fi} is the ith component's fiber density; and n is the number of blended/mixed components.

4.2.2.2 WEAVE-FACTOR ESTIMATION

The weave factor represents the number of warp and weft interlacements in a specific repeat, which equals average float, and is expressed as:

$$M = \frac{E}{I} \tag{4.5}$$

where; E is the number of threads per repeat; and I is the number of intersections per repeat of the cross-thread.

Considering some standard weaves, the values of warp and weft weave factors are presented in Table 4.1.

TABLE 4.1 Weave Factor for Standard Weaves

Standard Weaves	E_1	I_1	E_2	I_2	M_1	M_2
1/1 Plain	2	2	2	2	1	1
2/1 Twill	3	2	3	2	1.5	1.5
2/2 Warp rib	2	2	4	2	1	2
2/2 Weft rib	4	2	2	2	2	1

Note: E_1 and E_2 are the threads per repeat numbers in the warp and weft directions; I_1 and I_2 are the intersections per repeat numbers for the warp and weft threads.

- ***Example:*** Plain weave is represented as: $\frac{1}{\quad 1}$;

E_1 (ends per repeat numbers) = [1 + 1 = 2]; and I_2 (intersections per repeat numbers for warp yarn) = 1+ number of changes from up to down and vice versa = [1 + 1 = 2].

In the weave repetition, the number of intersections for each thread is not equal in various weaves. In such instances, the weave factor is calculated as follows:

$$M = \frac{\sum E}{\sum I} \tag{4.6}$$

4.3 RESULTS AND DISCUSSION

4.3.1 FOR PLAIN WEAVE FABRICS

The solution of p_2/D and h_1/D is achieved for a variety of warp crimp angles $[\theta_1]$ ranging from 0.1 to $\pi/2$ radians. Figure 4.2 displays one such relationship. The graphic is similar to what Peirce presented. It is a very useful fabric parameter connection for creating desired fabric structures. It is beneficial in the three cases listed below:

- Structures that are jammed;
- Fabrics that are not jammed;
- A special scenario with straight cross-threads.

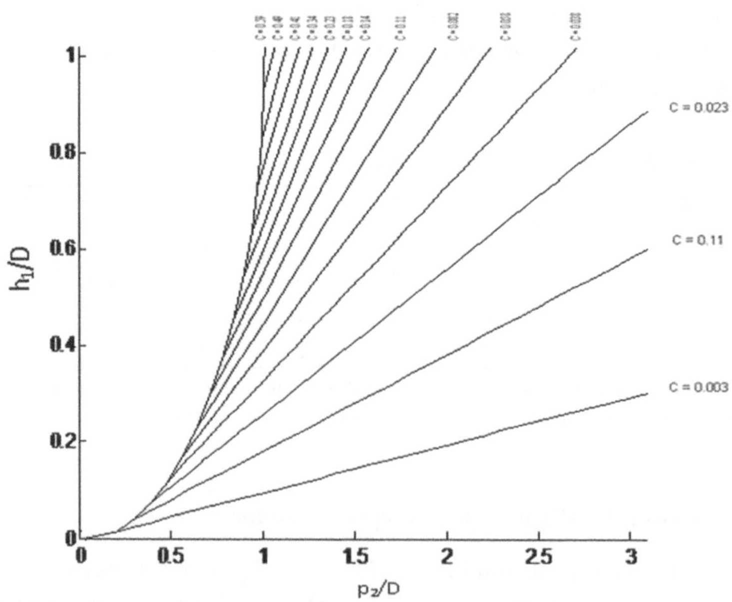

FIGURE 4.2 Relationship between thread spacing and crimp height.

4.3.1.1 STRUCTURES THAT ARE JAMMED

On the extreme left, Figure 4.2 depicts a nonlinear connection between the two FPs p and h. This curve, in reality, is in the warp direction for jamming. As can be observed, the jamming curve exhibits distinct p_2/D values as crimp increases. This curve, interestingly, is a portion of a circle, and its equation is:

$$\left(\frac{p_2}{D}\right)^2 + \left(\frac{h_1}{D} - 1\right)^2 = 1 \tag{4.7}$$

with the center at (0, 1) and radius equal to 1.

The parameters p_2/D and associated h_1/D for jamming in the fabric's warp direction can be determined from either this figure or the preceding equation.

The relationship between FPs over the full domain of structure being jammed in both directions may be found using the algorithm in Figure 4.3. It may be noted that the yarn spacing (p) and crimp amplitude (h) depend upon both yarn diameters and weave angle.

$$p_2 = D\sin\theta_1 = D\sin\frac{l_1}{D}$$

$$h_1 = D(1 - \cos\theta_1) = D\left(1 - \cos\frac{l_1}{D}\right) \tag{4.8}$$

It is known that $(h_1 + h_2) = (d_1 + d_2) = D$, or in other words, $\left[\frac{h_2}{D} = 1 - \frac{h_1}{D}\right]$
From equations of Peirce,

$$p_1 = D\sin\theta_2 = D\sin\frac{l_2}{D}$$

$$h_2 = D(1 - \cos\theta_2) = D\left(1 - \cos\frac{l_2}{D}\right) \tag{4.9}$$

The foregoing computation yielded the link between many relevant FPs, which are depicted in Figures 4.4–4.8. The link between p_2/D and p_1/D is seen in Figure 4.4. This graph demonstrates that the link between these factors is less sensitive at the two extremes. The connection is more sensitive in the p/D range toward 1. This sensitive range, in reality, corresponds to maximum crimp in one direction only. The range of change in p_2 and p_1 is 0.6 to 0.9. Beyond this range, thread spacing can only be changed in one direction.

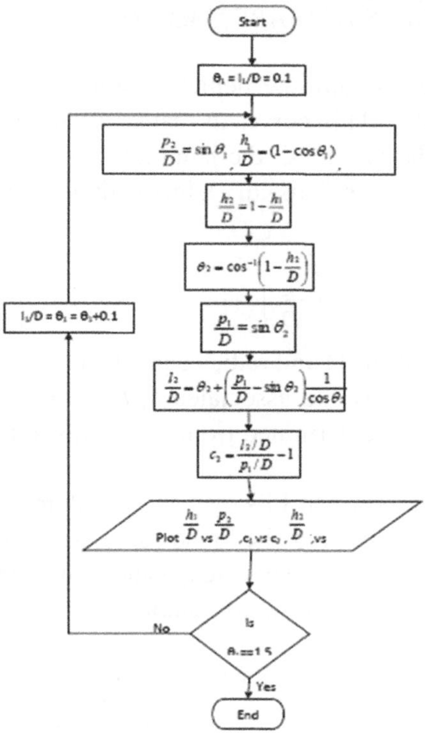

FIGURE 4.3 The flowchart for FPs in jammed structures.

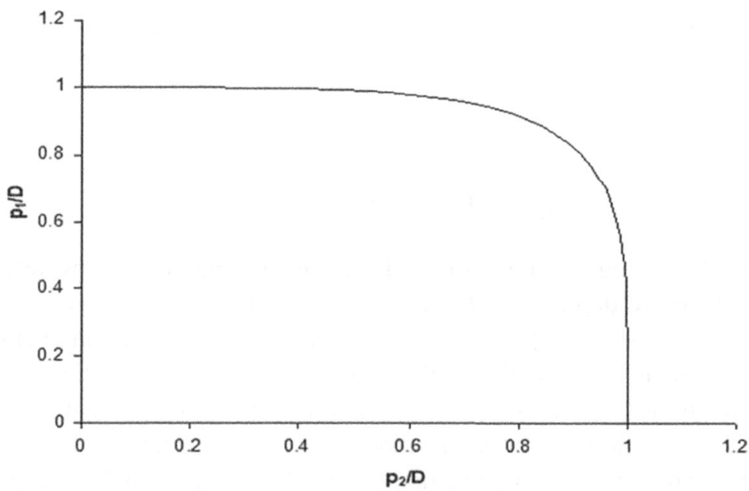

FIGURE 4.4 Warp and weft thread-spacing relationship for jammed fabrics.

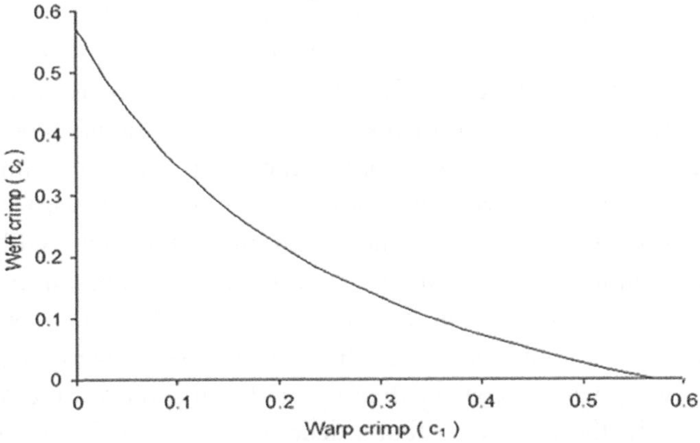

FIGURE 4.5 Fraction warp and weft crimp relationship for jammed fabrics.

Figure 4.5 depicts a helpful link between the crimps in both directions. It denotes a nonlinear inverse relation between c_1 and c_2. Maximum crimp values with zero crimp in the cross direction are provided by the axis intercepts. The cross threads are straight in this fabric's structure, while the intersecting threads do all of the bending. At either end, a little crimp change in one direction may allow for a large crimp movement in the opposite direction. The relationship between h_1/p_2 and h_2/p_1 is seen in Figure 4.6. Except at the two extremes, there is linearity between them. The behavior is actually a relationship between the square root of the crimp in two directions, with the constant being constant for any jammed fabrics. Additional sensible relationships are found between the warp and weft cover factors, as well as the fabric's cover factor and fabric mass (gsm).

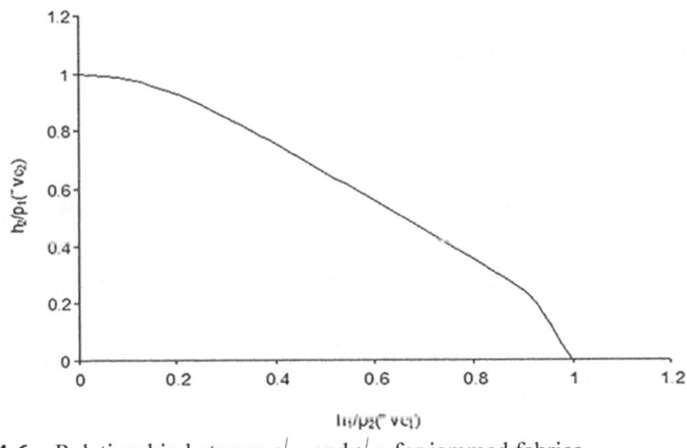

FIGURE 4.6 Relationship between $\sqrt{c_1}$ and $\sqrt{c_2}$ for jammed fabrics.

Figure 4.7 depicts the relationship between warp and weft cover factor for various weft-to-warp yarn diameter ratios (β). For all values of β, the relationship between the cover factors in two directions is only sensitive in a restricted range. For jammed structures, the relationship between the cover factors in two directions is interdependent. The maximum number of threads in the warp/weft direction is determined by yarn count and weaves. Maximum threads in one direction of the cloth will produce distinct maximum threads in the other direction. A change in the β value generates a noticeable shift in the curve. It is found that a coarser yarn in one direction in comparison to the other aids in improving the cover factor. When the warp yarn is coarser than the weft yarn, the warp cover factor increases while the weft cover factor decreases for $\beta = 0.5$. A similar effect may be seen with $\beta = 2$, when the weft yarn is coarser than the warp yarn.

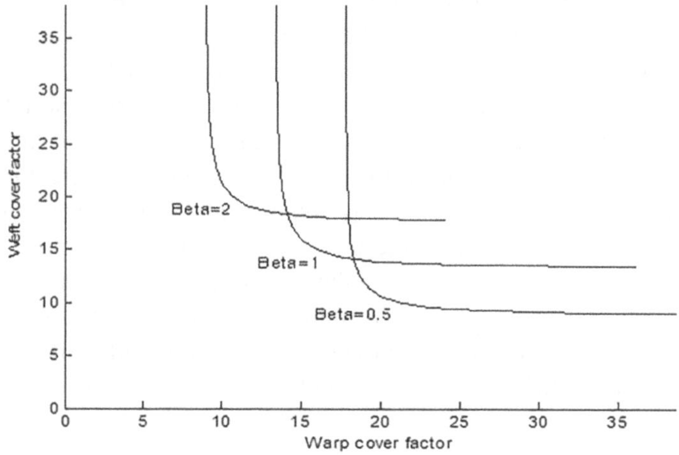

FIGURE 4.7 Relationship between warp and weft cover factor for different (β) of jammed fabrics.

Figure 4.8 depicts the positively linear relationship ($K_1 + K_2$) between fabric mass in (gsm) and cloth cover. The pattern may look self-explanatory. Increases in fabric mass and cloth cover factor for jammed fabrics may be achieved in a variety of methods, including maximum crimp in weft and zero crimp in warp; maximum crimp in warp and zero crimp in weft; and equal or unequal crimp in both directions, respectively. This explanation can be appreciated in reference to the nonlinear curve portion in Figure 4.2.

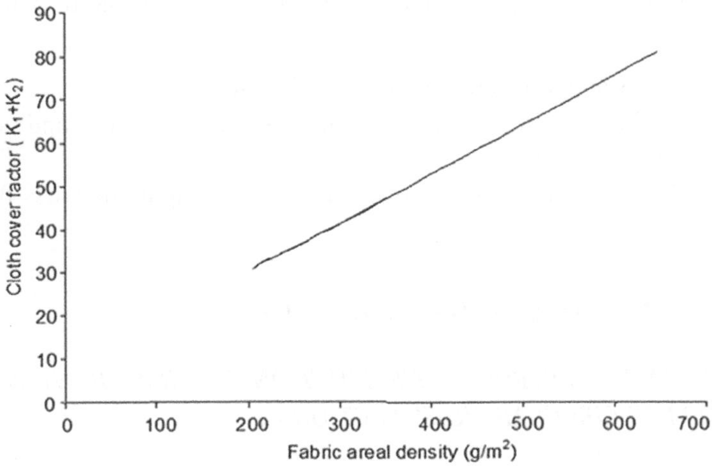

FIGURE 4.8 Fabric's cover factor and fabric mass relationship for jammed structures.

4.3.1.2 FABRICS THAT ARE NOT JAMMED

The relationship can be shown to be linear between p_2/D and the related h_1/D for various crimp levels. This connection is important for designing non-jammed structures for a wide range of crimp values. For any desired amount of warp crimp, the FPs may be determined using the aforementioned non-jammed linear relationship between p_2/D and h_1/D. Then h_2/D may be calculated from $(1 - h_1/D)$, and for this value of h_2/D, the value of p_1/D for the required weft crimp can be calculated. Thus, for any desired value of p_2/D; picks per cm; warp and weft yarn tex; and warp and weft crimp, all FPs may be attained. To obtain all FPs, one may select any other four parameters. For the non-jammed fabrics, the approach shown in Figure 4.1 may be utilized to obtain many solutions.

4.3.1.3 STRAIGHT CROSS-THREADS

The intersection of the horizontal line corresponding to $[h_1/D = 1]$ in Figure 4.2 yields all potential structures ranging from reasonably open to jammed arrangements. In this context, $h_2 = 0$, and $h_1 = D$.

The fabric designers are given options for different fabric compositions. Jamming and other loose structures are examples of these alternatives.

Using the reasoning described above, it is also feasible to obtain FPs for:

- A fabric that is jammed in both directions;
- A fabric with maximal crimp in one direction and straight cross-threads;
- A fabric that is not jammed and has no crimp in the cross-threads.

4.3.2 FOR NON-PLAIN-WEAVE FABRICS

4.3.2.1 JAMMED STRUCTURE'S EQUATION IN TERMS OF WEAVE FACTOR FOR CIRCULAR CROSS-SECTIONS

The weave factor may be used to translate the influence of weaves on fabric properties. The jammed cloth's general equation is sought for circular cross-sections.

The thread spacing [P_{t_1}] for a non-plain weave per repeat is presented in Figure 4.9 and follows the equation below:

$$P_{t1} = I_2 p_1 + (E_1 - I_2) d_1 \tag{4.10}$$

$$\text{The average thread-spacing, } \overline{p_1} = \frac{I_2 p_1 + (E_1 - I_2) d_1}{E_1}$$

$$\frac{E_1 \overline{P_1}}{I_2} = p_1 + \left(\frac{E_1}{I_2} - 1\right) d_1$$

$$M_1 \overline{P_1} = p_1 + (M_1 - 1) d_1$$

$$\frac{p_1}{D} = M_1 \frac{\overline{P_1}}{D} - (M_1 - 1) \frac{d_1}{D}$$

$$\frac{p_1}{D} = M_1 \frac{\overline{P_1}}{D} - \frac{(M_1 - 1)}{1 + \beta} \tag{4.11}$$

where; $\beta = d_2/d_1$

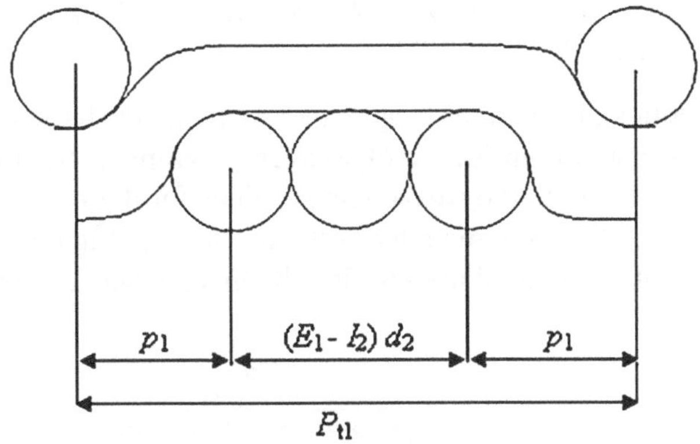

FIGURE 4.9 Jammed structure's configuration for 3/1 weave (along warp cross-section).

Similarly, by the interchange of suffix 1 and suffix 2, we get:

$$\frac{p_2}{D} = M_2 \frac{\overline{P_2}}{D} - (M_2 - 1)\frac{d_2}{D}$$

$$\frac{p_2}{D} = M_2 \frac{\overline{P_2}}{D} - (M_2 - 1)\frac{\beta}{1+\beta} \tag{4.12}$$

The following equation is acceptable for jammed fabrics:

$$\sqrt{1-\left(\frac{p_1}{D}\right)^2} + \sqrt{1-\left(\frac{p_2}{D}\right)^2} = 1$$

$$\sqrt{1-\left(M_1\frac{\overline{P_1}}{D} - \frac{(M_1-1)}{1+\beta}\right)^2} + \sqrt{1-\left(M_2\frac{\overline{P_2}}{D} - \frac{(M_2-1)\beta}{1+\beta}\right)^2} = 1$$

This equation is simply translated into warp and weft cover factor (K_1 and K_2) terms:

$$\sqrt{1-\left[\left(\frac{28.02\sqrt{\phi \tilde{n}_f} M_1}{K_1} - (M_1 - 1)\right)\frac{1}{1+\hat{a}}\right]^2} +$$

$$\sqrt{1-\left[\left(\frac{28.02\sqrt{\phi \tilde{n}_f} M_2}{K_2} - (M_2 - 1)\right)\frac{\hat{a}}{1+\hat{a}}\right]^2} = 1 \tag{4.13}$$

4.3.2.2 RELATIONSHIP BETWEEN FPS FOR CIRCULAR CROSS-SECTIONS OF VARIOUS WEAVES

The aforementioned equations for plain, satin, twill, and basket weaves are used to investigate the influence of weaves in jammed structures. Based on average float length, the weave factor values for these weaves include 1, 1.5, 2, and 2.5 for the remainder of the discussion. Figure 4.10 depicts the flowchart of the method used to find the linkage between various FPs.

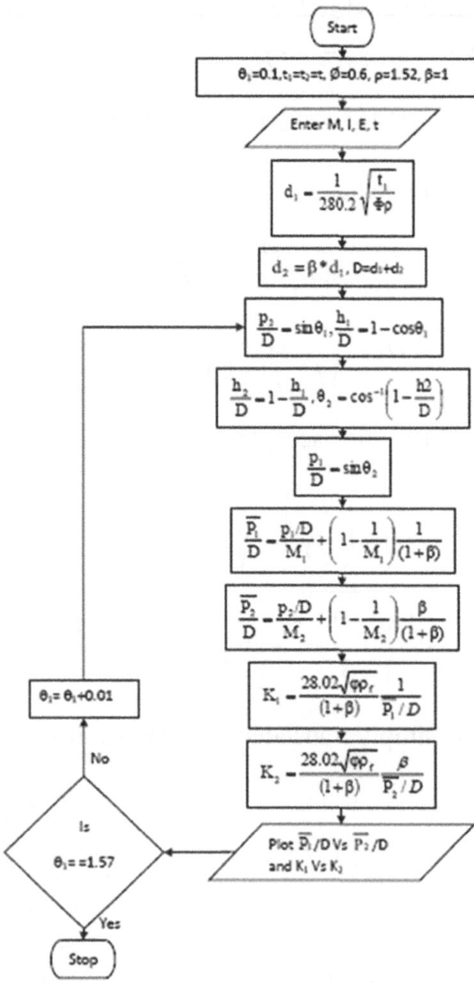

FIGURE 4.10 Flowchart for jammed structures to find the thread spacing and cover factor in circular cross-sections.

The relationship between p_{1avg}/D and p_{2avg}/D is seen in Figure 4.11. It can be observed that when float length increases, the curve's sensitivity diminishes in general. The range of p_1/D and p_2/D values is also decreased. This means that a weave with a greater float length has less structural flexibility.

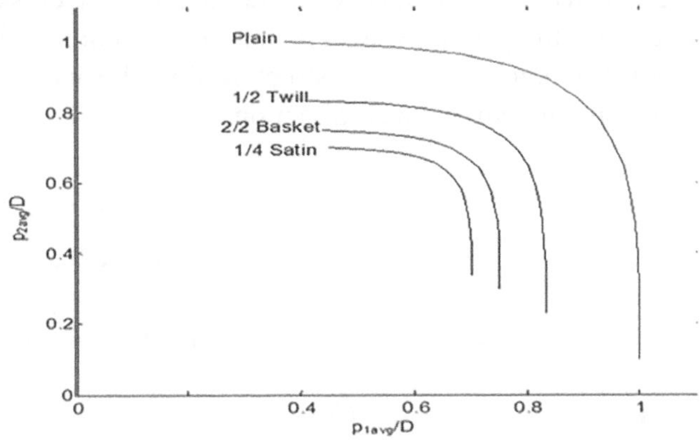

FIGURE 4.11 Average thread spacing in warp and weft relationship for jammed fabrics (circular cross-section, yarn tex = 30, $\rho = 1.52$, $\varnothing = 0.6$).

The warp and weft cover factor's relationship for a circular cross-section is shown in Figure 4.12. It is worth noting that the behavior is consistent among weaves. However, when float increases, the curve flips to greater values of weft cover factor. It should be noted that the behavior depicted in this picture is for virtual textiles. A jammed structure in actual fabrics is unlikely to keep a circular cross-section.

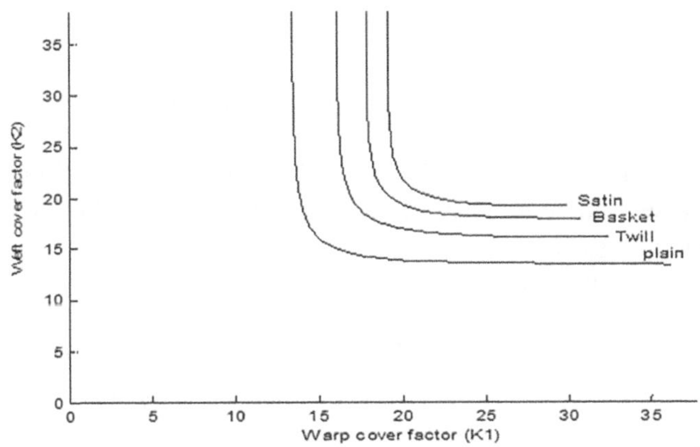

FIGURE 4.12 Warp and weft cover factor's relationship for jammed fabrics (circular cross-section, yarn tex = 30, $\rho = 1.52$, $\varnothing = 0.6$).

4.3.2.3 JAMMED STRUCTURE'S EQUATION FOR RACETRACK CROSS-SECTIONS IN TERMS OF WEAVE FACTOR

The yarn's cross-sections cannot remain circular in jammed fabrics, and the cross-sections will be altered. By considering racetrack cross-sections, it is simple to alter the geometry for circular cross-sections. Figure 4.13 depicts the construction of a jammed structure for a 1/3 weave racetrack cross-section along the fabric's weft direction.

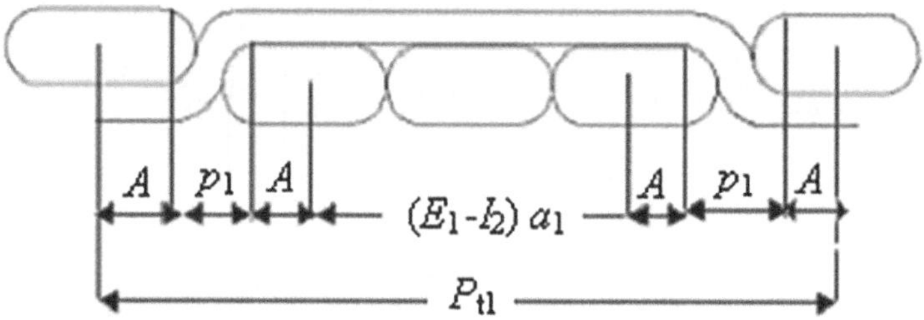

FIGURE 4.13 Jammed structure's configuration for 3/1 weave (along warp cross-section).

$$\text{Here, } A = \left(\frac{a_1 - b_1}{2}\right)$$

The thread spacing (P_{t_1}) for a non-plain weave per repeat is as follows:

$$P_{t1} = I_2 p_1 + (E_1 - I_2) a_1 + 4\left(\frac{a_1 - b_1}{2}\right) \tag{4.14}$$

Similarly,

$$\text{similarly, } P_{t2} = I_1 p_2 + (E_2 - I_1) a_2 + 4\left(\frac{a_2 - b_2}{2}\right) \tag{4.15}$$

where p_1 and p_2 are horizontal spacing between the semi-circular threads in the intersection zone. Here, a and b are the major and minor diameters of racetrack cross-sections.

The average thread spacing,

$$\text{The average thread spacing, } \overline{P_1} = \frac{p_1}{M_1} + \left(1 - \frac{1}{M_1}\right) a_1 + \frac{4}{M_1 I_2}\left(\frac{a_1 - b_1}{2}\right) \tag{4.16}$$

Similarly,

$$\overline{P_2}=\frac{p_2}{M_2}+\left(1-\frac{1}{M_2}\right)a_2+\frac{4}{M_2I_1}\left(\frac{a_2-b_2}{2}\right) \tag{4.17}$$

As a result, circular thread geometry analysis may be performed on the intersection zone of the racetrack cross-sections.

$$L_1=4\left(\frac{a_2-b_2}{2}\right)+(E_2-I_1)a_2+I_1\times l_1 \tag{4.18}$$

In the fabric, the total warp crimp is as follows:

$$C_1=\frac{L_1}{P_{t2}}-1$$

From the jamming considerations, p_1 and p_2 can be calculated for the circular thread geometry by the use of the following:

$$\sqrt{1-\left(\frac{p_1}{B}\right)^2}+\sqrt{1-\left(\frac{p_2}{B}\right)^2}=1$$

It is important to remember that [p/B] refers to the semi-circular portion of the racetrack cross-section and is equivalent to [p/D] for the circular cross-section. As a result, the p/D ratio values may be utilized for p/B.

$$\sqrt{1-\left(M_1\frac{\overline{P_1}}{B}-\frac{4}{BI_2}\left(\frac{a_1-b_1}{2}\right)-(M_1-1)\frac{a_1}{B}\right)^2}+$$
$$\sqrt{1-\left(M_2\frac{\overline{P_2}}{B}-\frac{4}{BI_1}\left(\frac{a_2-b_2}{2}\right)-(M_2-1)\frac{a_2}{B}\right)^2}=1$$

This equation may be reduced to the following forms:

$$\sqrt{1-\left(M_1\frac{\overline{P_1}}{B}-\frac{2(1-e)}{e(1+\beta)I_2}-\frac{(M_1-1)}{e(1+\beta)}\right)^2}+$$
$$\sqrt{1-\left(M_2\frac{\overline{P_2}}{B}-\frac{2(1-e)\beta}{e(1+\beta)I_1}-\frac{(M_2-1)}{e(1+\beta)}\right)^2}=1$$

It is assumed that $\{e_1 = e_2 = e\}$,
where; $[e = b/a]$

The above equation may be simply translated into warp and weft cover factor as follows:

$$\sqrt{1-\left(\frac{28.02\sqrt{\varphi \tilde{n}_f}M_1}{(1+\hat{a})K_1}\sqrt{1+\frac{4}{\pi}\left(\frac{1}{e}-1\right)}-\frac{2(1-e)}{e(1+\hat{a})I}-\frac{(M_1-1)}{e(1+\hat{a})}\right)^2}$$

$$+\sqrt{1-\left(\frac{28.02\sqrt{\varphi \tilde{n}_f}M_2\hat{a}}{(1+\hat{a})K_2}\sqrt{1+\frac{4}{\pi}\left(\frac{1}{e}-1\right)}-\frac{2(1-e)}{e(1+\hat{a})I}-\frac{(M_2-1)}{e(1+\hat{a})}\right)^2}=1 \quad (4.19)$$

4.3.2.4 RELATIONSHIP BETWEEN FPS IN RACETRACK CROSS-SECTIONS

Figure 4.14 depicts the algorithm's flowchart used to compute thread spacing and cover factor in jammed racetrack cross-sections. The relation between FPs such as p_2 and p_1; p_1 and c_2 for the jammed racetrack cross-section is investigated. It illustrates that the behavior of these two parameters is similar to that of the circular cross-section, although it moves toward greater thread spacing values and describes the behavior of true fabrics.

Figures 4.15 and 4.16 depict the warp and weft cover factor's relationship for various weaves. As previously noted, actual or true fabrics have unique weave variances, as opposed to circular cross-sections. The scope of cover factors is reduced as float length increases. Moreover, crimp and fabric cover may be calculated using the aforementioned two equations, as well as:

$$d^2 = b^2\left[\frac{\pi}{4}+\left(\frac{1}{e}-1\right)\right] \text{ and } \frac{b_2}{b_1}=\hat{a} \quad (4.20)$$

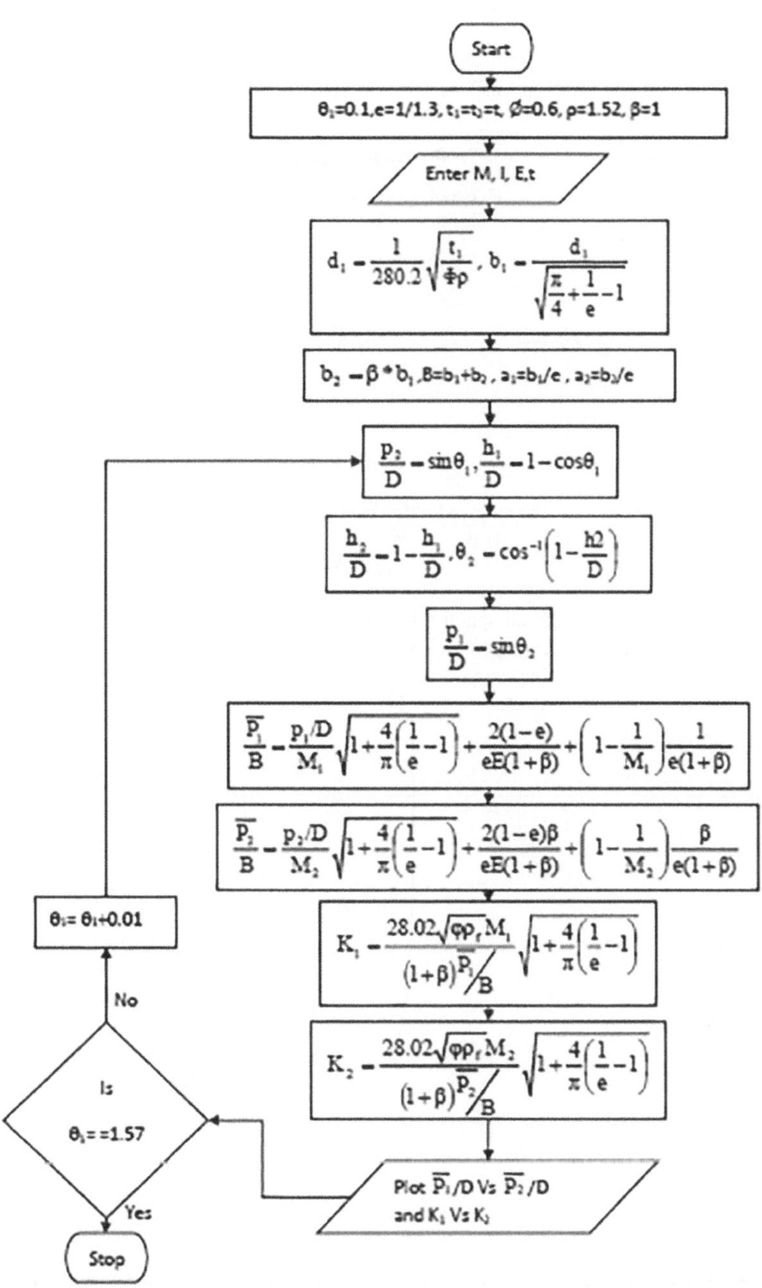

FIGURE 4.14 Jammed structure's flowchart for determining thread spacing and cover factor in racetrack cross-sections.

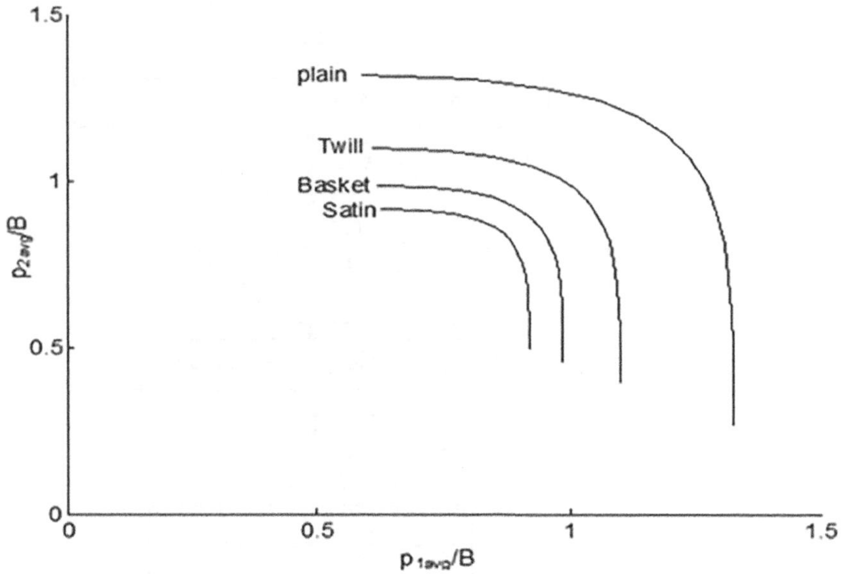

FIGURE 4.15 Average thread spacing in warp and weft relationship for jammed fabrics (racetrack cross-sections, yarn tex = 30, $\beta = 1$, $\rho = 1.52$, $\varnothing = 0.6$).

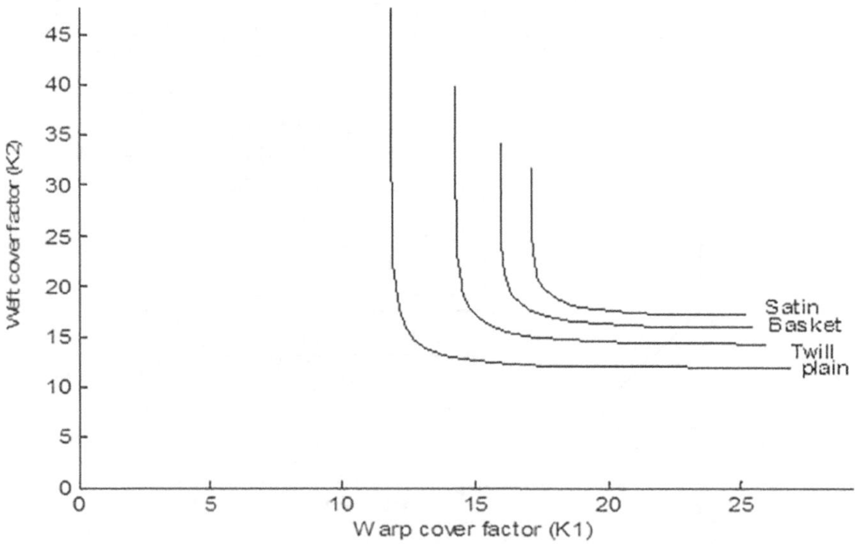

FIGURE 4.16 Warp and weft cover factor's relationship for jammed fabrics (racetrack cross-sections, yarn tex = 30, $\beta = 1$, $\rho = 1.52$, $\varnothing = 0.6$).

4.4 CONCLUSION

Soft computing has fully utilized the Peirce fabric geometrical model for plain as well as non-plain weaves. The algorithm may be used to obtain the interrelationships between different FPs demonstrated for jammed structures. It is envisaged that the ability to pick FPs for any end application with desired fabric qualities would assist fabric designers.

Furthermore, soft computing's capability is proven for non-plain weave materials. The maximum threads per centimeter of a yarn spun from any fiber and the fabric generated from any weave may be predicted. The connection between cover factors in the warp and weft direction is shown for circular and racetrack cross-sections of plain, satin, twill, and basket weaves. Non-plain weave fabrics provide more flexibility for expanding fabric bulk and cover. As a result, they broaden the reach of fabric designers.

KEYWORDS

- **circular cross-section**
- **cover-factor**
- **fabric-designers**
- **fabric-parameters**
- **non-plain weaves**
- **Peirce fabric geometrical model**
- **plain weaves**
- **warp and weft**

REFERENCES

Backer, S. (1952). The mechanics of bent yarns. *Textile Research Journal*, *22*(10), 668–681.

Chou, T. W. (1992). Microstructural design of fiber composites. *NASA STI/Recon Technical Report A*, *92*, 50452.

Dickson, J. B. (1954). Practical loom experience on weavability limits. *Textile Research Journal*, *24*(12), 1083–1093.

Grosberg, P., & Kedia, S. (1966). The mechanical properties of woven fabrics: Part I: The initial load extension modulus of woven fabrics. *Textile Research Journal*, *36*(1), 71–79.

Guénon, V. A. (1987). *Interlaminar fracture toughness of a three-dimensional composite* (Doctoral dissertation, University of Delaware).

Guénon, V. A., Chou, T. W., & Gillespie, J. W. (1989). Toughness properties of a three-dimensional carbon-epoxy composite. *Journal of materials science, 24*, 4168–4175.

Hearle, J. W. S., & Amirbayat, J. (1988). The design of a multipurpose fabric tester. *Journal of the Textile Institute, 79*(4), 588–597.

http://www.i-designer.co.kr, i-Designer (accessed on 7th June 2025).

http://www.texclub.com, Texpro (accessed on 7th June 2025).

Hu, J. (2004). *Structure and Mechanics of Woven Fabrics*. Elsevier.

Kawabata, S., Postle, R., & Niwa, M. (Eds.). (1982). *Objective Specification of Fabric Quality, Mechanical Properties and Performance*. Textile Machinery Society of Japan.

Kemp, A. (1958). *An extension of Peirce's cloth geometry to the treatment of non-circular threads.* T44–T48.

Kim, H. A., & Kim, S. J. (2010). Simulation of the weave structural design of synthetic woven fabrics. *Fibers and Polymers, 11*, 905–910.

Kim, S. J. (2007). Preliminary data-base system for the fabric structural design. *Journal of Textile Engineering, 53*(5), 159–171.

Ko, F. K. (1989). Three-dimensional fabrics for composites. *Elsevier Science Publishers, Textile Structural Composites,*, 129–171.

Kuwabara, S. (1986). Effects of solubility parameters on disperse dyeing of poly (ethylene terephthalate) films. *Sen'i Gakkaishi, 42*(4), T240–T247.

Lindberg, J., Behre, B., & Dahlberg, B. (1961). Part III: Shearing and buckling of various commercial fabrics. *Textile Research Journal, 31*(2), 99–122.

Ly, N. G., Tester, D. H., Buckenham, P., Roczniok, A. F., Adriaansen, A. L., Scaysbrook, F., & De Jong, S. (1991). Simple instruments for quality control by finishers and tailors. *Textile Research Journal, 61*(7), 402–406.

Mallick, P. K. (2007). *Fiber-Reinforced Composites: Materials, Manufacturing, and Design*. CRC Press.

Peirce, F. T. (1937). The geometry of cloth structure. *Journal of the Textile Institute Transactions, 28*(3), T45–T96.

Postle, R., Carnaby, G. A., & De Jong, S. (1988). The mechanics of wool structures. *(No Title)*.

Seyam, A. M., & Harrison, P. W. (2002). The Structural Design of Woven Fabrics: Theory and Practice-A critical appreciation of recent developments. *Textile Progress, 31*(3).

Tanzawa, Y., Watanabe, N., & Ishikawa, T. (1997). Interlaminar delamination toughness and strength of 3D orthogonal interlocked fabric composite. In *Proceedings of the Eleventh International Conference on Composite MaterialsGold Coast, Australia, 14–18 July* (pp. 47–57).

Tanzawa, Y., Watanabe, N., & Ishikawa, T. (1999). Interlaminar fracture toughness of 3-D orthogonal interlocked fabric composites. *Composites Science and Technology, 59*(8), 1261–1270.

CHAPTER 5

A Critical Analysis of the Use of Machine Learning in the Field of Reliability Prediction

GUAMBO MONDO, JOHN KAMIT, AND KAMALAKANTA MUDULI*

Papua New Guinea University of Technology, Lae, Morobe Province, Papua New Guinea

*Corresponding author

ABSTRACT

This study is based on quality assurance and control of quality through machine learning. The massive data age today brings along its own vulnerabilities and errors as well. This signifies the need to study the impact of machine learning and how quality assurance can be achieved with the help of machine learning, specifically how it contributes to quality assurance and quality control in industry—for example, equipment maintenance, processes in manufacturing, determining the remaining useful life of parts, and so forth. The study does not deal with quality assurance of machine learning itself, but rather how machine learning is used to ascertain quality assurance and reliability to guarantee standards (specification limits) and customer satisfaction in mass production, manufacturing, marketing, transport, weather forecasts, and so forth.

5.1 INTRODUCTION

5.1.1 BACKGROUND OF MACHINE LEARNING

The transformation of existing raw data into usable data is known as machine learning. The study of inferring from training data to extract

Advances in Industrial Manufacturing and Materials: Integrating a New Era of Computational Modeling and Machine Learning. Debesh Mishra and Swagatika Mishra (Eds.)

DOI: 10.1201/9781779640475-5

important and pertinent information is based on or uses adaptive algorithms. As described by Kung (2014) and Ouerhani et al. (2022), it is a successful data-driven strategy for intelligent categorization and prediction. Fundamentally, machine learning, an application of artificial intelligence, allows software programs to accurately forecast outcomes or identify patterns using data sets that have been gathered or "mined."

5.1.2 THE PROBLEM SCENARIO IN MACHINE LEARNING QUALITY ASSURANCE AND QUALITY CONTROL

The current data age brings along massive data sets that are used in numerous applications. Moreover, this massive data comes with varying degrees and potential for errors associated with the data used in applications or stages of applications. In essence, the data in machine learning is stored, mined, extracted, and inferred for machine learning, producing inferred output as "trained" (Figure 5.1).

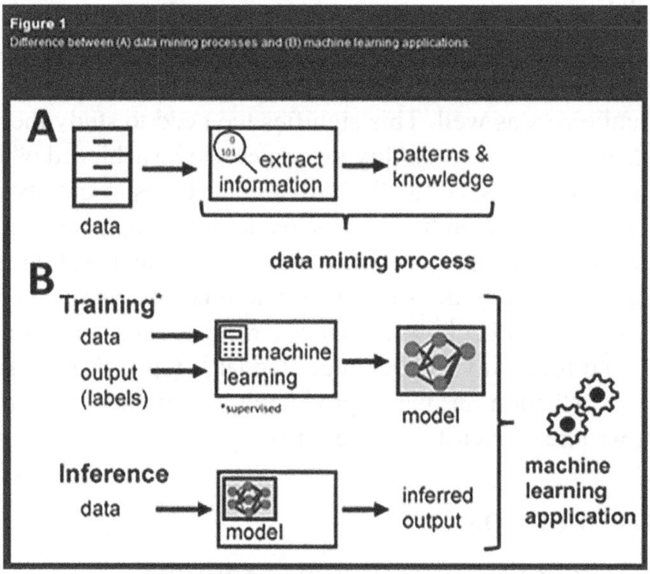

FIGURE 5.1 Machine learning background.

Machine learning faces challenges of all sorts, from the use of material properties to computational artificial inferences in predicting outcomes. These challenges have to be met individually because of the complexity and nature of the quality of predictions anticipated.

5.1.3 *MACHINE LEARNING QUALITY PREDICTIONS*

5.1.3.1 MACHINE LEARNING OVERVIEW

Machine learning comes in different categories by understanding different aspects of what data is collected and relating different features to one another: to determine relationships, anomalies, and predict the former from the latter.

Machine learning categories and subcategories or tasks are shown in Figure 5.2, where the latter are not meant to be exhaustive. Both supervised and semi-supervised learning modes contain anomaly detection algorithms; the more widely used are in unsupervised mode. Research on semi-supervised regression is less extensive than that on its classification equivalent.

5.1.3.2 ACTIVE MACHINE LEARNING

Not shown in Figure 5.2, machine learning is active learning; for example, in which the algorithm can query the user to provide labels to carefully selected previously unlabeled data. In this manner, machine learning is active learning.

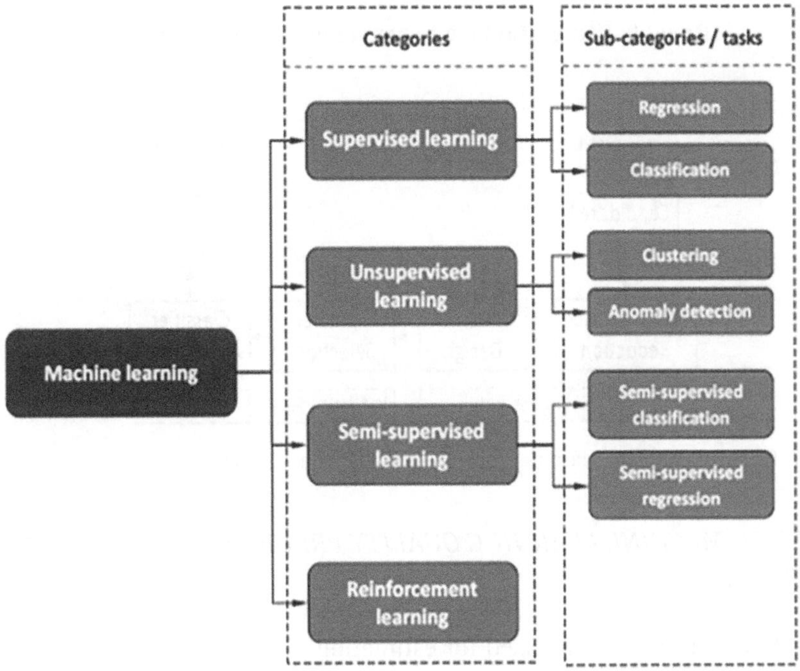

FIGURE 5.2 Categories of machine learning.

5.1.3.3 MACHINE LEARNING'S BENEFITS FOR QUALITY ASSURANCE

The benefits of quality assurance brought about by machine learning are increasing as more and more applications rely on machine learning. These benefits include an increase in the broad categories of customer satisfaction, quicker sales and marketing times, an increase in production and profitability, detection of high risks and early warnings in risk management, prediction of occurrences, and so forth.

When customers are happy and stay happy, firms are happy and profitable, and machine learning plays an increasing role in these instances and in other marketplaces and industries. Machine learning also eases the challenges imposed by rigorous and awkward designs by taking over tasks that can be risky for humans. Machines can be easily replaced, so in high-risk and inaccessible areas, they can be deployed instead of risking humans. Machine learning can evaluate outcomes sooner and convey results more easily to stakeholders, thus staging higher degrees of reliability, quality, output, early warning, accessibility, and ease of maintenance. Machine learning's benefits for quality assurance are shown in Figure 5.3.

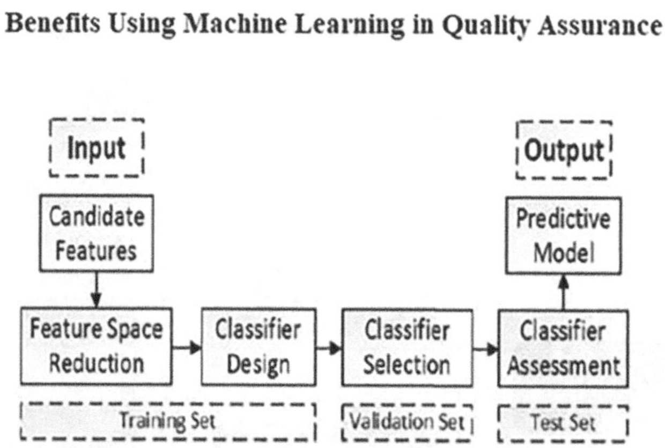

FIGURE 5.3 Machine learning's benefits for quality assurance.

5.1.3.4 MACHINE LEARNING QUALITY PREDICTIONS OF GENERAL NATURES

Machine learning can be used for estimating:

- The remaining useful life of assets;

- Anomaly and fault detection;
- Health monitoring;
- Maintenance planning; and
- Degradation assessment.

The objective is generally to tease out novel, more precise dataset findings for improved reliability and safety-informed decision-making in system design and operations, and more effective accident prevention (Xu and Saleh, 2021).

5.1.3.5 MACHINE LEARNING ALGORITHMS

The basics of machine learning are (Xu and Saleh, 2021):

- Datasets for training and testing purposes;
- An objective function or loss function to optimize, for example, a sum of squared errors or a likelihood function; and
- An optimization algorithm and a model for the data (e.g., linear, nonlinear, nonparametric).

5.2 LITERATURE REVIEW

5.2.1 INTRODUCTION TO MACHINE LEARNING LITERATURE REVIEW

Although many industrial and commercial machines and systems offer compelling quality assurance, a balance of performance, sound reliability, high-performance control, and reduced overall system costs, possible faults due to defects, broken components, and misalignment may affect the quality of the product and cause unexpected breakdowns, injuries, or even death to workers or users (Nikfar et al., 2022). This brings about the crucial need for real-time condition monitoring (CM) and preventive maintenance (PvM) tools to be employed to detect anomalies and predict patterns of possible defects in process plants and equipment, ultimately determining actions to avoid machine failures, damage, and unsafe occurrences.

In the "Industry 4.0" concept, maintenance has emerged as a crucial field with significant emphasis and impact on the uptime and efficiency of production systems, costs, safety, and improvements.

This fourth industrial revolution (Industry 4.0), however, relies on the big data age with the opportunity to transform industries through artificial intelligence concepts, by integrating physical and digital systems to create machine learning. This machine learning ultimately is utilized to predict outcomes from datasets and assist with the "human-machine interface (HMI)" without human "touch" (Ouerhani et al., 2022).

5.2.2 WHY MACHINE LEARNING IN MAINTENANCE?

Any business can attribute revenue, efficiency, and safety to maintenance as a business process. The goal of maintenance is to increase equipment reliability, speed up production, and lower costs. It is described as the company's procedure for keeping equipment in a desirable state of efficient operation (Kung, 2014).

Preventive, corrective, and predictive maintenance (PdM) are the most common subcategories of maintenance management (Ouerhani et al., 2022). Corrective maintenance is sometimes known as run-to-failure maintenance. Corrective maintenance is reactive and sometimes takes place ad hoc. It involves the replacement or repair of an asset when a defect occurs. It typically happens following the failure of an aviation asset, and the goal is to return it to operational status as soon as feasible, either by repair or replacement.

Machine learning focuses on computer algorithm creation, and its main aim is to make it possible for computer systems (or rather systems) to "learn" without human involvement instantaneously (Surya, 2019). Different machine learning algorithms can be trained, tested, and used to explore CM and PvM techniques to monitor, detect, predict, and avoid failures. The health and conditions of machines and components are monitored by CM and preventive maintenance (PM). CM and PM are used interchangeably, but the main difference is time. CM focuses on real-time, while PvM focuses more on early detection of abnormalities.

5.2.3 CONDITION MONITORING (CM)

Early predictions and awareness of deviations in equipment limitations and process conditions are essential for the health of production and the quality assurance of products. These predictions prevent product recalls, breakdowns, production shutdowns, hefty maintenance downtime, long time between repairs, and so forth. Knowing the conditions and status of equipment and processes beforehand, with the help of machine learning, saves huge costs in losses of time, production, recalls, and so forth.

5.2.4 PREVENTIVE MAINTENANCE (PVM)

Similar to CM, PM can be assisted by machine learning by alerting or providing early warnings of deviations from acceptable time between maintenance, preventing breakdowns. This raises the alarm for PvM prior to breakdown, thus maintaining quality assurance, sound status, and preventing costly maintenance and repair (Wan et al., 2017).

5.2.5 PREDICTIVE MAINTENANCE (PDM)

The fundamental tenet of PvM is the performance of maintenance duties in accordance with the manufacturer's specifications. To reduce malfunctions and maintain the proper application of equipment, it is constrained by manufacturer deadlines, and spare parts are replaced or equipment components based on the operating handbook/procedures. This type of maintenance is carried out while the machinery is operational and is planned to ensure that all necessary resources are available. PdM, then, is a way of thinking or doing that employs the equipment and systems of a plant in their real operational state to improve plant operations and/or processes. PdM uses analytical techniques and sensor technology to foresee potential equipment problems (Eddarhri et al., 2022; Herrero and Zorrilla, 2022; Nikfar et al., 2022; Wan et al., 2017).

5.2.6 TYPES OF MACHINE LEARNING METHODS IN QUALITY CONTROL

The two important pillars of machine learning are neural networks and kernel methods. Neural methods have been dominant (Kung, 2014), while kernel methods are now gaining momentum in this big data age. However, quality assurance in machine learning is fundamentally based on neural methods. For quality control, machine learning uses neural networks, more so in foreseen predictions and sensor technology to stay ahead of faults, breakdowns, and critical failures.

5.2.7 MACHINE LEARNING TECHNIQUES

In general, the properties of materials can be studied or tested to predict their behavior. In most cases, these properties are subjected to conditions or treatments that produce predicted patterns. Some of these material properties are:

- Composition of the material (pure or compounds);
- Amount of material(s);
- Electrical properties;
- Thermal heat properties;
- Strength of materials;
- Treatment of materials.

We will look at case examples to see how machine learning is employed in total quality assurance.

5.2.8 MACHINE LEARNING IS THE FUTURE OF QUALITY ASSURANCE

In general, machine learning models are an example of a data framework that uses data from a certain set to infer assumptions about a new observation. To analyze the present dataset and forecast current trends, machine learning algorithms are created.

A collection of organized algorithms updated through the process of learning makes up neural networks. The learning process requires data inputs in order to produce outputs and then compare them to predetermined outcomes. Additionally, programs use technology to identify data trends and understand the massive amount of data to create quick and reliable performance.

Artificial intelligence technology is used in machine learning to give programs the ability to understand dynamically without the use of scripts or direct human input. Experience-based applications find their usefulness in this manner.

5.2.9 MACHINE LEARNING BASED FOR RELIABILITY ENGINEERING AND SAFETY

It can be hard to navigate and integrate the existing, extensive but disjointed literature on machine learning for reliability and safety applications. By offering a summary of, a roadmap for, and emphasizing the key landmarks and pathways in this constantly expanding analytical terrain, we facilitate this task in this work.

First, we give a general overview of the various machine learning categories, subcategories, or tasks, and then include a few models and algorithms that relate to each category. The employment of machine learning in dependability and safety applications is then examined in the past. Several publications are reviewed in each category and subcategory and provide a brief discussion on deep learning to draw attention to its rising application and unique benefits (Alsina et al., 2018).

5.2.10 MACHINE LEARNING BASED ON ELECTRICAL PROPERTIES

Machine learning utilizes many properties in nature to predict and perform outputs. For example, the application of machine learning algorithms in quality assurance of the fermentation process of black tea is based on the electrical properties of tea leaves. Clearly stating the use of machine learning in tea leaves' electrical properties can create quality standards/ assurance and determine the quality of fermented black tea by the contents of catechins and tea pigment in tea leaves (Zhu et al., 2019).

5.2.11 MACHINE LEARNING IN THE AVIATION INDUSTRY

Like all industries, aeronautical equipment upkeep requires regular, expensive work. Preventive and corrective maintenance, however, is the most frequently used type of maintenance in the field of aviation. Aeronautical wiring companies generate a large amount of data that, with appropriate processing and intelligent systems, can provide useful information for the reduction of machine downtime. As a result, PdM is implemented, with the goal of predicting system failure by continuously monitoring its status and planning and performing maintenance actions in advance (Eddarhri et al., 2022). Aeronautical wire businesses that utilize preventative and corrective maintenance could benefit from the advantages of artificial intelligence outlined in this study, using machine learning and data mining approaches.

5.2.12 DATA-DRIVEN INDUSTRIAL CASE STUDIES USING MACHINE LEARNING TECHNIQUES TO FORECAST COMPONENT RELIABILITY

One of the key contributions in machine learning study is the comparison of a broad and diversified range of machine learning models to suit the dependability of a set of industrial real situations, given the significance and advantages of machine learning approaches for reliability prediction. These examples were drawn from various settings and actual industrial parts. From the initial data collection stage to the final performance evaluation of the machine learning models to forecast the component dependability, the application of machine learning models can show the failure prediction process (Alsina et al., 2018).

The major goal is to determine if there is a single data-driven learning model that can outperform other models under various reliability prediction scenarios.

5.2.12.1 UTILIZING MACHINE LEARNING TO ESTIMATE THE LIFESPAN OF SPARE PARTS

The majority of industries choose a hybrid maintenance approach that combines corrective (R2F) and preventive (PvM) strategies, with the former being applied when a breakdown occurs and no PvM is scheduled. These two tactics, however, do have disadvantages. Industries that practice R2F maintenance frequently postpone maintenance tasks while accepting the risk that their assets may not be available. In the case of PvM maintenance, it could result in the replacement of spare parts that are still in good condition. Adopting PdM, which has the potential to identify a problem before it happens, is one alternative. However, PdM is not a practical choice for many sectors (Macedo et al., 2022).

5.2.12.2 MACHINE LEARNING MODEL FOR DOSE OPTIMIZATION IN CT QUALITY ASSURANCE: A METHOD FOR ESTIMATING MATERIAL POTENTIAL

In a retrospective, cross-sectional study, the application of machine learning to identify CT scans with the potential for dosage optimization for quality assurance is evaluated. To predict the volumetric computed tomography dose index (CTDI) based on scan patient data (scanner, study protocol, patient age, sex, and water-equivalent diameter [Dw]), the CT chest exams model (of 3199 CT patients) was trained and optimized in this case example. Performance was measured by calculating the RMSE (root-mean-squared error). For CT quality assurance, a difference of 18.3% between the expected and observed CTDI appears to be sufficient (Meineke et al., 2019).

5.3 METHODS

It is best to consider two examples of methodologies used in predictive quality assurance.

5.3.1 METHODOLOGIES

Methodologies used in machine learning predictions differ from case to case fundamentally. The best way to describe machine learning predictions is to apply machine learning to a particular case and describe its learned behavioral patterns as a result of its output behavior individually.

The methodologies herein are individually set out from different studies to show what is meant in these circumstances.

5.3.1.1 MACHINE LEARNING – FUTURE OF QUALITY ASSURANCE

Between "April 2004 and May 2021," a significant automotive multimedia assembly firm gathered 118,776 records of maintenance requests combined with spare parts movements performed. The information is a collection of maintenance orders for the equipment's replacement of spare components. Additionally, each equipment machine's type of maintenance, the technician who carried it out, and the item that was changed could all be registered.

Depending on the variety of components needed for the maintenance, the precise sequence may also be linked to multiple records (e.g., maintenance that requires three separate spare parts may be encapsulated in three records, corresponding to each part). Whenever the part moves again, the order may likewise be reopened. Furthermore, there are six different kinds of motions related to spare parts: stock entries in the supplier's labyrinth, supplier returns of the parts, shortages of stock for a maintenance order, and supplier returns of the component to the maze. Maintenance orders come in three varieties: preventative, corrective, and improvement (which involves making adjustments to the equipment to bring about improvements).

5.3.1.2 ESTIMATION OF THE LIFESPAN OF SPARE PARTS USING MACHINE LEARNING APPROACH

The majority of data-driven (remaining useful-life [RUL]) studies use machine learning RUL prediction models that are fed data from equipment measurements (e.g., image, temperature levels, and equipment's operating time). In one study, a somewhat different RUL machine learning prediction approach was proposed, whereby the lifetime of spare parts is predicted using historical records of corrective maintenance (which are less difficult to obtain).

Specifically, the lifespan is expressed as the total number of units generated over two successive corrective maintenance activities. It is regarded as a current dataset for case studies, comprising 18,355 records with RUL measurements taken from an automobile multimedia assembly firm. In the event that a regression task modeling is used, eight different machine learning algorithms are examined and contrasted: random forest (RF), decision tree (DT), XGBoost (XB), extra trees (ET), histogram-based gradient boosting machine (HGBM), light gradient boost machine (LGBM), a Gaussian kernel support vector machine (SVM), and linear support vector machine (LSVM). To handle inputs with higher cardinality of different levels, two categorical pre-processing approaches were used: Inverse document frequency (IDF) and percentage categorical pruned (PCP). Additionally, the effectiveness of the machine learning prediction models was assessed by assuming a realistic "Rolling Window" method, which replicates several training and testing runs throughout time.

Ultimately, an eXplainable Artificial Intelligence (XAI) technique is used to quantify the influence of the accepted industrial inputs in the RUL predictions, allowing for a more thorough analysis of the best RUL prediction model in terms of its retrieved information.

5.3.2 DATA ANALYSIS

As with methodologies in each individual case, the data analysis also differs across cases, including data retrieval and storage methods. Nonetheless, the application algorithms forming models cannot be exhausted. Many models can be created. The data logging centers and analysis play equally essential roles in the stages of machine learning.

Normally, data logging is essential in all individual cases. Data logging may be regarded as the process of gathering data, storing it, and displaying one or more datasets to analyze activities and assist in predicting future events. This is the foundation of machine learning in reality.

Looking at each case, we see a similar activity pattern of data logging and dataset display in prediction and outputs.

5.4 RECOMMENDATIONS

In general, the future of machine learning is here to stay as more and more datasets are available for public consumption, making machine learning more reliable and readily available in designing and setting new platforms for predicting and analyzing behavioral outputs.

It is highly recommended that the prediction tools and datasets that constitute the backbone of machine learning or HMIs be accepted into industry without obvious restrictions so that designs based on these tools do not limit future reliance on machine learning to solve any problem.

5.5 CONCLUSION

Machine learning may remain a great helping hand in human history if utilized in an ethical manner. Machine learning has transformed the industry and is heading into Industry 5.0, with little known to limit the use of machine learning.

5.5.1 LIMITATIONS AND SCOPE OF THE FUTURE OF MACHINE LEARNING QUALITY CONTROL

In general, the future of machine learning is here to stay as more and more industries are progressing into Industry 5.0. There are robots and machines handling what we term "daily activities" that were previously performed by humans in quality control. It is an era created through data logging and dataset analysis to predict events and responses as if done by humans. Safety has improved by the use of machine learning, replacing human risk and interference, and supposedly creating reliability with greater precision, among other benefits.

Not that machine learning is replacing humans, but HMIs have made difficult jobs achievable, more reliably done without compromising safety, and performed in harsh or high-risk environments where humans cannot even reach, nor have previously thought possible. Repeated tasks can be achieved precisely, and corrections can be made instantaneously to improve quality, safety, production, and create more reliable future developments. Machine learning has paved the way for even more in-depth studies and future hopes for greater dimensions in quality control and assurance.

Most limitations fall in the likelihood of personal preferences rather than known and available prediction tools collectively done and stored in large corporate organizations. For example, machine learning datasets for a particular food flavor may not be available for public consumption, with the intention of protecting future sales and marketing (as in the case of Coca-Cola). The flavored drink keeps its formula highly secret, and thus these datasets are secret, as we see it—limitations that can be seen in machine learning where datasets are not made publicly available, or, as in the case of GPS, to protect public privacy, and so forth.

Apart from other setbacks, the staging platforms are vulnerable to attacks, such as hacking and introducing virus attacks on algorithms. Otherwise, machine learning is great for humankind.

KEYWORDS

- **corrective maintenance**
- **decision-tree**
- **machine learning**
- **quality assurance**
- **quality control**
- **random forest**
- **spare parts lifetime**
- **volumetric computed tomography dose index**

REFERENCES

Alsina, E. F., Chica, M., Trawiński, K., & Regattieri, A. (2018). On the use of machine learning methods to predict component reliability from data-driven industrial case studies. *The International Journal of Advanced Manufacturing Technology*, *94*, 2419–2433.

Eddarhri, M., Adib, J., Hain, M., & Marzak, A. (2022). Towards predictive maintenance: the case of the aeronautical industry. *Procedia Computer Science*, *203*, 769–774.

Herrero, R. D., & Zorrilla, M. (2022). An I4. 0 data intensive platform suitable for the deployment of machine learning models: a predictive maintenance service case study. *Procedia Computer Science*, *200*, 1014–1023.

Kung, S. Y. (2014). *Kernel Methods and Machine Learning*. Cambridge University Press.

Macedo, L., Matos, L. M., Cortez, P., Domingues, A., Moreira, G., & Pilastri, A. L. (2022, February). A Machine Learning Approach for Spare Parts Lifetime Estimation. In *ICAART (3)* (pp. 765–772).

Meineke, A., Rubbert, C., Sawicki, L. M., Thomas, C., Klosterkemper, Y., Appel, E., Caspers, J., Bethge, O. T., Kröpil, P., Antoch, G., & Boos, J. (2019). Potential of a machine-learning model for dose optimization in CT quality assurance. *European radiology*, *29*, 3705–3713.

Nikfar, M., Bitencourt, J., & Mykoniatis, K. (2022). A two-phase machine learning approach for predictive maintenance of low voltage industrial motors. *Procedia Computer Science*, *200*, 111–120.

Ouerhani, N., Loehr, B., Rizzotti-Kaddouri, A., Santo De Pinho, D., Limat, A., & Schinderholz, P. (2022). Data-driven thermal deviation prediction in turning machine-tool-a comparative analysis of machine learning algorithms. *Procedia Computer Science*, *200*, 185–193.

Surya, L. (2019). Machine learning-future of quality assurance. *International Journal of Emerging Technologies and Innovative Research (www. jetir. org)*, ISSN 2349–5162.

Wan, J., Tang, S., Li, D., Wang, S., Liu, C., Abbas, H., & Vasilakos, A. V. (2017). A manufacturing big data solution for active preventive maintenance. *IEEE Transactions on Industrial Informatics*, *13*(4), 2039–2047.

Xu, Z., & Saleh, J. H. (2021). Machine learning for reliability engineering and safety applications: Review of current status and future opportunities. *Reliability Engineering & System Safety*, *211*, 107530.

Zhu, H., Liu, F., Ye, Y., Chen, L., Liu, J., Gui, A., Zhang, J., & Dong, C. (2019). Application of machine learning algorithms in quality assurance of fermentation process of black tea--based on electrical properties. *Journal of Food Engineering*, *263*, 165–172.

CHAPTER 6

Intrinsic Aspects of Fiber-Matrix Interface in FRP Composites: Special Focus on Utility

S. BEURA,[1,*] U. K. MOHANTY,[1,2] S. N. PATI,[1] and P. K. SATAPATHY[1]

[1]*Department of Mechanical Engineering, ITER, SOA (Deemed to be) University, Bhubaneswar, Odisha, India*

[2]*Department of Metallurgical and Materials Engineering, NIT Rourkela, Odisha, India*

*Corresponding author

ABSTRACT

Fiber-matrix wettability, as influenced by fiber-matrix compatibility, interlaminar shear strength, and the possible presence of any residual stresses at the interface, are important factors inherent to any fiber-reinforced polymer composite. These factors dictate the structural stability of the composite. By definition, any composite is a heterogeneous material in which the structural stability is ensured by a good fiber-matrix interfacial bond. Good interfacial bond strength enhances the durability, reliability, and integrity of the composite. On the other hand, poor interfacial bond strength may not be adequate to resist the deteriorating effects of the environment, resulting in stress accumulation and propagation, which can lead to the material's failure. The write-up in this case is an attempt to evaluate these degrading factors responsible for lowering the mechanical performance of an FRP composite.

6.1 INTRODUCTION

The fiber-matrix interface in FRP composites is characterized by its ability to transfer any external load from the weak matrix to the strong,

Advances in Industrial Manufacturing and Materials. Integrating a New Era of Computational Modeling and Machine Learning. Debesh Mishra and Swagatika Mishra (Eds.)

DOI: 10.1201/9781779640475-6

load-bearing reinforcement fiber. Its bond strength pertains to the tightness and intimacy between the fiber and the matrix, representing the durability, reliability, and integrity of the FRP composite. A weak interfacial bond fails to resist the degrading effects of the environment, such as humidity and temperature fluctuations, when stress accumulation and propagation adversely affect the structural stability of the material. Certain intrinsic aspects of the fiber-matrix interface, including fiber-matrix wettability concerns, interfacial bond strength issues, and the presence of any residual stresses at the interface, need to be analyzed to assess the effects of such factors on the mechanical performance of the FRP composite. The effects of these factors on the mechanical properties of the material are discussed in detail.

Certain very-welcome specific properties of fiber-reinforced-polymeric composites, which include light weight, good insulating properties, high specific strength, corrosion resistance, and toughness, including improved fatigue properties, have made the material a natural choice for many specific uses. Even in many instances, the FRP composites have replaced their metallic counterparts. The selection of the matrix-fiber combination, the various processes adopted for fabrication, different treatments for enhancing fiber-matrix wettability and adhesion, and so on, offer a great deal of possibilities to engineer the material for specific uses. Indeed, the material has conquered air, land, and marine fronts concerning its value-added use as aerospace materials, material of choice in automotive industries, ship-building, and civil construction. Thus, the versatility of FRP composite materials is beyond any doubt. To appreciate and achieve the full potential application of the material, one must have an adequate knowledge of the salient features of the material. In this respect, the fiber-matrix interface, which governs stress accumulation and propagation in the FRP composite and reflects its structural integrity, deserves a special mention (Yoo et al., 2017). An appropriate interfacial bond ensures good shear strength, toughness, and stiffness properties of the composite (Chakraverty et al., 2020a; Rafiee and Mazhari, 2016), necessitating full understanding of the intricacies of the material (Koyanagi et al., 2014). Many experiments have been conducted to assess the interfacial properties, adopting single fiber composite specimens (Carloni et al., 2017; Chen and Chouw, 2018; Fuentes et al., 2015; Kwon et al., 2019; Li et al., 2019a, 2020; Prusty et al., 2017; Shi et al., 2021; Shin et al., 2017; Tan et al., 2018; Zhou et al., 2017), and bulk composite specimens (Barczewski et al., 2016; Beura et al.,

2021a; Egoshi et al., 2020; Fitriah et al., 2017; Goh et al., 2018; Jeyakumar et al., 2017; Liu et al., 2020; Zhao et al., 2018). Though the test procedures adopted in these test methods are relatively simple, the data interpretation for judging detailed behavior of the interface and hence that of the material pose real challenges. The test results pertaining to pull-out and push-out tests for interfacial bond-strength assessments can be cited as an example toward this end (Møder et al., 2000). The result of the aforementioned test shows higher bond strength at the interface obtained by the pull-out test as compared to the push-out test, which establishes the dependence of the interfacial bond strength on the loading pattern, after a cumbersome interpretation of the test readings. Additionally, one must appreciate the importance of "work of adhesion" (W_a), interfacial shear stress (τ), and the energy release rate (G_{i_c}) in deciding the bond strength at the interface, as well as its stability and structural integrity. In this regard, the fundamental acid-base interactions, covalent bonds, secondary or van der Waals forces, and dipolar interactions between the fiber and the matrix must be well understood. These form the basic processes of fiber-matrix interfacial bond formation. Good fiber-matrix wettability enhances the fiber-matrix bond formation, ensuring a strong adhesion between the two. Poor wettability results in the formation of voids at the interface, which can lead to stress concentrations, ultimately lowering the shear strength of the composite as a whole. Analysis of the wetting parameters can assist in conceiving and developing proper conditions for any specific use of the material.

The interfacial residual stresses constitute important considerations for a healthy and stable interface. While factors like mechanical interlocking, the chemical nature of the fiber and the matrix, molecular characteristics, and atomic arrangements decide the interfacial bond strength (Chakraverty et al., 2020b) in isolation and/or combination, the existence of residual stresses at the interface puts the material in a pre-loaded stressed condition. The tensile residual stresses in the direction of the fibers, in particular, severely affect the structural integrity of the FRP composite. However, compressive residual stresses in the transverse direction favorably contribute to the prevailing lower adhesion between the fiber and the matrix (Mohamed et al., 2020; Zhang et al., 2016).

Keeping the above in mind, the present article attempts to highlight and analyze the intrinsic aspects of interface characteristics. The wettability issues, bond strength, and residual stress concerns in particular are presented and discussed in detail.

6.1.1 ATTRIBUTES OF THE INTERFACE

6.1.1.1 WETTABILITY

Wettability refers to the ability of the liquid polymer matrix to wet the solid fiber reinforcement in an FRP composite. This is instrumental in establishing good fiber-matrix adhesion. A low contact angle between the matrix and the fiber fosters good wettability between the two, which is necessary but not a sufficient condition for good adhesion and strong bonding between the two. This is because even excellent wettability may be accompanied by weak van der Waals-type, low-energy bonds. In addition to the importance of a low contact angle, the ability of the liquid matrix to wet the solid fiber also depends on time and temperature of contact, interfacial reactions, stoichiometry, surface roughness and geometry, heat of formation, and the electronic configuration of the fiber and matrix. In a polymer matrix composite, the polymer matrix must wet and penetrate the fiber tows. Incomplete wetting may give rise to interfacial flaws in the form of voids. The stress concentrations induced by these voids result in a lowering of the interfacial shear stress (IFSS).

The wetting parameters include work of adhesion, fiber surface energy, and interfacial energy, besides the contact angle generated when the liquid polymer and the solid fiber are in contact (Lu et al., 2019). Thus, fiber-matrix wettability can be analyzed on the basis of these parameters. Such an analysis would assist in evaluating the interface in terms of its humidity resistance and interfacial adhesion (Luner and Oh, 2001; Park et al., 2002). This paves the way for engineering the interface and tailoring composites with improved mechanical performance to suit any specific end use.

The equilibrium between three phases—the solid, the liquid, and the vapor—when in contact, is represented by Young's equation. The equation is given as (Chen et al., 2006):

$$\gamma_l \cos\theta = \gamma_s - \gamma_{sl} \tag{6.1}$$

where; θ is the contact angle between the solid (e.g., the fiber) and the liquid (e.g., the matrix) phases; γ_l is the experimentally determined surface tension of the liquid polymer; γ_s is the surface free energy of the solid fiber; and γ_{sl} is the interfacial energy between the fiber and the matrix.

Clearly, when θ assumes lower values and tends to zero, good wettability of the fiber by the matrix is ensured. Under these conditions, *cosθ*, and hence γ_s, the fiber surface free energy, is high. In other words, a low contact angle between the fiber and the polymer matrix and the high surface free energy of the solid fiber ensure good wettability. Under these conditions, the fiber can be well impregnated by the matrix liquid, ensuring better fiber–matrix adhesion. Also, a high surface free energy is indicative of the presence of more polar groups on the fiber surface, being responsible for a strong interaction between the polar groups and the resin matrix (Luner and Oh, 2001; Park et al., 2002). The result is an improved fiber–matrix adhesion that enhances the IFSS of the composite.

Chen et al. (2006) determined the contact angles for glass, carbon, and aramid fibers separately with different testing liquids, adopting a Cahn dynamic contact angles analysis system. From the respective contact angle values, the total surface free energy was calculated and split into the respective polar and dispersive components using Owens–Wendt equations given below (Hoecker and Karger-Kocsis, 1996; Lee and Lee, 2002; Luner and Oh, 2001; Park et al., 2002, 2003; Saïhi et al., 2002; Wu, 2004):

$$\gamma_l(1+\cos\theta) = 2(\gamma_s^p \gamma_l^p)^{1/2} + 2(\gamma_s^d \gamma_l^d)^{1/2} \tag{6.2}$$

$$\gamma_{total} = \gamma_s^p + \gamma_s^d \tag{6.3}$$

where; θ is the dynamic contact angle between the testing liquid and the respective fibers; γ_l is the surface tension of the testing liquid; γ_{total} is the surface free energy of the respective fibers; and γ_s^p, γ_s^d is the respective polar and dispersive components.

The calculated θ values were the highest for the glass fibers, decreasing for the carbon and aramid fibers in that order. Also, the surface free energy values were found to be the highest for the carbon fibers, decreasing for the aramid and the glass fibers in that order. With this information and the concept that a high surface free energy would enhance the wettability of the fiber, as a consequence of which the fiber would adhere tightly to the resin, Chen et al. (2006) fabricated fiber–PPESK thermoplastic composites with glass, aramid, and carbon fibers (the PPESK thermoplastic represents a novel, processed polymer with excellent thermal performance; T_g = 287°C) to examine the above. The wetabilities of the three types of fibers were, obviously, different. Glass fibers, with the highest θ values and the lowest surface free energy, were not well impregnated with the PPESK polymer solution, exhibiting poor fiber–matrix adhesion.

"W_a," the thermodynamic work of adhesion, was also considered as a measure of adhesion between the fiber and the matrix. "W_a" is defined as the work to be performed to separate the fiber and matrix phases reversibly from their interfaces in the equilibrium state. "W_a" in terms of Dupre's equation, can be given as (Chen et al., 2006):

$$W_a = 2[(\gamma_s^d \gamma_l^d)^{1/2} + (\gamma_s^p \gamma_l^p)^{1/2}] \tag{6.4}$$

where; the terms have their usual meanings.

When the surface free energy values are higher, as in the case of carbon fibers, the "W_a" value is also higher. This is indicative of a stronger fiber–matrix adhesion.

It was, thus, established theoretically that the wettability of the carbon fiber was the best, followed by the aramid and the glass fibers. The same was also established through an investigation conducted by the researchers with fiber–PPESK composites.

6.1.1.2 MOISTURE ABSORPTION TREND

The moisture absorption trend of the composite of the carbon, aramid, and glass fiber, as reported by the researchers, is presented in Figure 6.1 (Chen et al., 2006). Moisture is found to permeate into the glass-fiber–PPESK

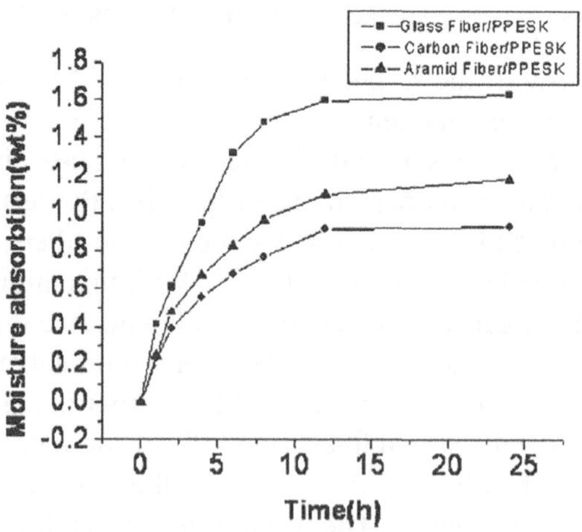

FIGURE 6.1 Moisture absorption of the composites.

Source: Reprinted with permission from Chen et al., 2006. Copyright 2006. John Wiley and Sons

composite at the highest rate and to the highest extent. The water permeation to the composite takes place through the interface, as the matrix and the fiber have low water absorption capabilities. These glass fiber–PPESK composites do not exhibit good fiber–matrix adhesion, which indirectly reflects the highest extents of interfacial flaws (Kumosa et al., 2004). This trend is also true for the carbon fiber and aramid fiber–PPESK composites. These composites show good adhesion and lesser extents of interfacial flaws, respectively, as compared to the glass fiber–impregnated PPESK composites.

6.1.1.3 HUMID RESISTANCE

The mechanical property variations of the composites after immersion in boiling water at 100C for 48 hrs. are presented in Figure 6.2 (Chen et al., 2006). It is evident that there are only slight changes in mechanical properties. This means, for the respective composites, the humid resistance properties are excellent. However, major deteriorations in the ILSS values can be observed, the deterioration being the highest for the glass fiber–PPESK composite, resulting in the lowest value being set for the glass fibers.

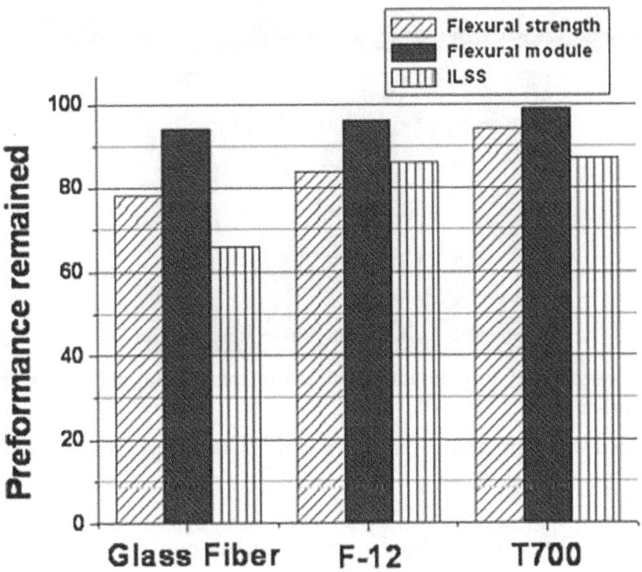

FIGURE 6.2 Humid resistance properties of the composites.

Source: Reprinted with permission from Chen et al., 2006. Copyright 2006. John Wiley and Sons

6.1.1.4 SEM FRACTOGRAPHS

The SEM fractographs of the treated composites are presented in Figure 6.3 (Chen et al., 2006). In the glass fiber–PPESK composites, with the minimum exhibited wettability of the glass fibers, it is seen that the fibers are loosely held by the matrix polymer. This reflects poor fiber–matrix adhesion and maximum interfacial flaws. The clear fiber surfaces in the concerned fractographs clearly indicate the above. On the other hand, for the carbon and aramid fibers, the fractographs reveal considerable matrix deformation along with the breakage of the fibers. For these fibers, the good wettability results in better fiber–matrix adhesions compared to the glass fiber–PPESK composite.

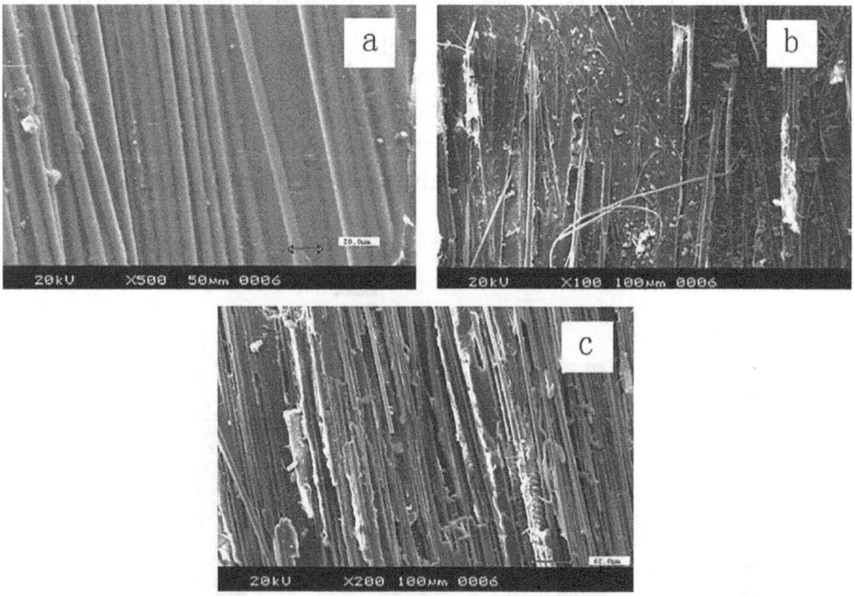

FIGURE 6.3 Morphology of interlaminar shear ruptures for: (a) glass fiber; (b) aramid fiber; and (c) carbon fiber.

Source: Reprinted with permission from Chen et al., 2006. Copyright 2006. John Wiley and Sons

Thus, wettability, which has a strong influence on the fiber–matrix adhesion, can be engineered. This can be accomplished by:

i. Selection of the fiber–matrix combination with the best compatibility.

ii. Fiber surface chemistry modifications for the optimization of fiber–matrix adhesion, as revealed by the analysis of fiber surface free energy.

Wettability, therefore, is an important tool for tailoring the interface for the specific end use of the concerned composite, by conveniently appropriating its mechanical performance.

6.1.1.5 INTERFACIAL BONDING

Interfacial bonding in FRP composites is as good or as bad as the fiber–matrix adhesion at the interface, which is a direct function of the properties of the fiber depending on its morphology and the ability of the fiber and the matrix to diffuse into each other. In addition, interfacial adhesion is also attributed to the abilities of the matrix and the fiber to adhere/cling to each other and their wetting characteristics. It also depends on the electrostatic attraction between the constituent phases, reaction bonding, and exchange reaction bonding (reaction bonding following the formation of a new compound). To be more specific, the interfacial bond strength is judged by the quality of interfacial adhesion (Chakraverty et al., 2017, 2020b; Ismael et al., 2020; Qasem et al., 2020).

Interfacial bonds can be formed by molecular entanglements, electrostatic attractions, interdiffusion of elements, chemical reaction bonding, exchange reaction bonding (as in MMCs), and mechanical interlocking. This is illustrated in Figure 6.4. In brief, chemical and molecular characteristics, atomic arrangements, and surface topography of the fiber—which determine the fiber–matrix adhesion strength—are also important considerations for establishing appreciable interfacial bond strength (Chakraverty et al., 2020b, c). Besides the thermodynamical wetting characteristics that greatly influence the fiber–matrix physical attractions, the adhesion strength leading to interfacial bond strength also depends on relatively stronger acid–base interactions and/or covalent bonds. Also, in addition to the principal mechanisms discussed above, hydrogen bonding, van der Waals forces, and other low-energy forces play a role in interfacial bond formation. The interfacial region is open to the actions of all these mechanisms, in isolation and/or in combination, to produce the final interfacial bond.

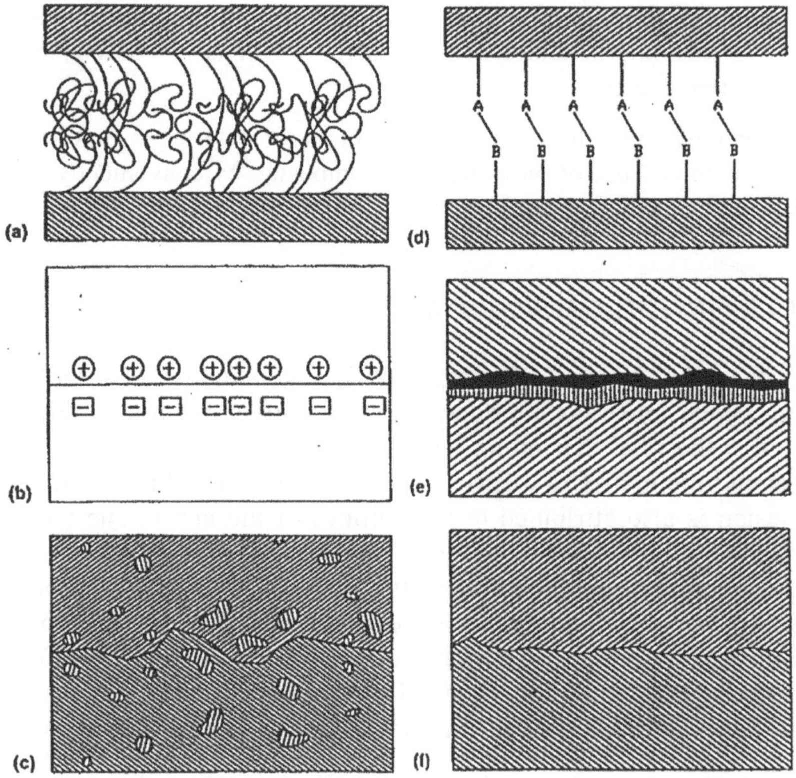

FIGURE 6.4 Interface bonds formed: (a) by molecular entanglement; (b) by electrostatic attraction; (c) by interdiffusion of elements; (d) by chemical reaction between group A and group B on the other surface; (e) by chemical reaction following formation of a new compound(s), particularly in MMCs; (f) by mechanical interlocking.

Source: Reprinted with permission from Zhou et al., 2016. Copyright 2016. Elsevier.

6.2 TYPES OF INTERFACIAL BONDS

The important possible bonding types at the interface can be categorized as physical bonding, chemical bonding, and mechanical bonding, respectively.

The final interfacial bonding is due to the impact of one or any combination of the above at any instant of time.

6.2.1 PHYSICAL BONDING

The physical attraction between the electrically neutral matrix and the fiber in an FRP composite, which determines the adhesion between the two, is largely dictated by the ability of the liquid matrix to wet the solid reinforcement. The fiber–matrix adhesion strength is characterized in terms of the reversible work of adhesion, W_a, and the measured adhesion strength is given by τ, which is the IFSS. W_a, the work of adhesion, is defined as the work required to break the unit area of any adhesion contact between the matrix and the fiber. W_a quantifies matrix–fiber wetting and is given as:

$$W_a = \gamma_l + \gamma_s - \gamma_{ls} \tag{6.5}$$

where γ_s and γ_l denote the surface free energies of the solid and the liquid phases; and γ_{ls} denotes that of the interface. Obviously, the physical bond strength is augmented when the wetting of the fiber by the matrix is improved. Hypothetically, the surface free energy γ of any phase has a dispersive nonpolar part and a polar interaction component, both contributing to the surface energy of the respective phases. Thus, $\gamma_l = \gamma_l^d + \gamma_l^p$ and $\gamma_s = \gamma_s^d + \gamma_s^p$ respectively denote the dispersive (nonpolar) and polar components. It is reported (Wu, 2017) that adhesion assumes the maximum value when $\gamma_s^p = \gamma_l^p$, that is, when the polar interactions of the constituent phases contributing to the surface tension of the solid and liquid phases are equal. Such a condition improves the interfacial physical bond.

The measured adhesion strength, τ, also indicates the force required for the failure of the fiber–matrix interface (Mittal, 2014; Packham, 1996; Sheu and Shyu, 1994). Thus, τ also characterizes the physical bond between the fiber and the matrix. This statement is supported by many researchers (Jacobasch et al., 1995; Mäder et al., 1994; Nardin and Schultz, 1993; Nardin et al., 1990). A weak relationship between τ and the work of adhesion, W_a, has been reported in the literature (Harding and Berg, 1997; Nardin et al., 1991; Ulkem and Schreiber, 1994), wherein τ is given as:

$$\tau = \frac{F}{\pi d l} \tag{6.6}$$

where F is the fiber pull-out force obtained from laboratory tests; d is the diameter of the fiber; and l is the contact length of the fiber.

Many researchers, however, have reported a quantitative relationship between τ and W_a (Jacobasch et al., 1992; Swolfs et al., 2016). Their observations could be a result of the application of coupling agents to the surface of the reinforcement fiber. Coupling agents, in general, decrease the surface energy of inorganic fibers like glass fiber and carbon fibers, resulting in a decrease in the W_a value. However, the adhesion strength, τ, increases when coupling agents are used, contradicting the W_a decrease. There is a lack of theoretical understanding to explain this contradiction. Also, when the fiber surface is treated with coupling agents like silane—specifically in the case of acidic or neutral reinforcements—an electrostatic charge difference between the reinforcement fiber and the matrix is established, which contributes to attraction bonding between the two (Figure 6.4(b)) (Hull, 1998). Physical bonding between the fiber and the matrix at the interface may also involve secondary or van der Waals forces, dipolar interactions, and hydrogen bonding. The bond energy in these cases, however, is very low.

6.2.2 CHEMICAL BONDING

The adhesive bond formed between two chemical groups—one on the fiber surface and the other in the matrix—both groups being chemically compatible, is a consequence of the chemical bonding mechanism. The bond is formed due to chemical reactions between the two groups, activated by the prevailing thermal conditions. Two polymer surfaces, on account of molecular entanglement (Figure 6.4(a)) (Hull, 1998), due to interdiffusion of the polymer-matrix molecules into the network of molecules on the fiber surface, also assist the bond formation. The ease of molecular motion, type of constituents involved, and spatial arrangement of atoms in the concerned molecules determine the extent of these interdiffusions and hence the tangled molecular bonds. The chemical bonds include all types of covalent, ionic, and metallic bonds and are very strong.

6.2.3 MECHANICAL BONDING

Mechanical bonding at the fiber–matrix interface is a consequence of mechanical interlocking on the fiber surface. Surface oxidation treatments of the fibers contribute immensely to mechanical interlocking in carbon fiber/polymer matrix composites, forming large numbers of pits, corrugations, and so on, on the carbon fiber surface (Hull, 1998). The strength of mechanical interlocking bonds is not very high in transverse tension. However, the longitudinal shear strength due to these bonds in the direction of the fibers may be very high, depending on the surface roughness of the fibers. In addition to the surface roughness of the fibers, which serves as a simple geometrical aspect of mechanical interlocking, internal stresses in the composite acting normal to the fiber directions may give rise to clamping stresses, adding to mechanical anchoring. These stresses could be the result of shrinkage of the matrix on curing and/or unequal thermal coefficients of expansion of the matrix polymer and the reinforcement fibers.

6.3 ASSESSMENT OF INTERFACIAL BOND STRENGTH

The strength of the bond at the interface should neither be very low nor very high. It should be optimum. Strong or weak interfacial bonds can be conceived by prioritizing the required response of the composite to its service conditions. A strong interface may be needed when shear, off-axis, and compression strengths of the composite are the focus. On the other hand, the ability of the composite to resist the propagation of cracks perpendicular to the fiber orientation is enhanced—allowing flexibility—when the bond strength is relatively weak. Thus, interfaces need to be optimized, which necessarily means that the composition and micromechanical properties of the interface must be well addressed, characterized, and controlled (Liu et al., 2015).

Simultaneously, it must be appreciated that there could be instances when too high an interfacial bond strength may decrease composite performance (Parveen et al., 2020). This again underscores the importance of understanding the necessity of measuring interface mechanical properties. One common example of such properties is the adhesion strength of the fiber–matrix combination at the interface. This can be measured by adopting different experimental techniques that describe a real material property (Beura et al., 2021b).

It is important to note that, under practical situations, structural load components are not aligned in the fiber direction only but may occur at various angles to the fiber axis. It must therefore be ensured that a maximum stress level is maintained across the interface—from fiber to matrix—without discrepancies. In this respect, the strain rate during characterization, which governs the relaxation time (low loading speeds allow for longer relaxation time), also constitutes an important consideration.

6.3.1 *METHODS FOR EVALUATING INTERFACIAL PROPERTIES*

Several methods involving a single fiber or a group of fibers is adopted for the measurement of interfacial adhesion. In this respect, it must be emphasized that although most of the tests are relatively simple, the experimental data interpretation is difficult and may require the utility of advanced models. In addition to predicting the fiber–matrix adhesion at the interface, an appropriate and just analysis of the data paves the way for understanding the fundamental physics and chemistry responsible for fiber–matrix adhesion.

It is important to note that comparison of data for assessing properties and understanding the details of fiber–matrix adhesion, when using different methods, may lead to serious disagreements and misunderstandings.

The variety of methods adopted for the measurement of adhesion can be grouped into three basic categories:

- Fiber pull-out method;
- Fiber fragmentation method; and
- Fiber micro-indentation method.

Detailed descriptions of these methods are beyond the scope of this chapter, as they are well-established in the field.

Out of all these methods, however, the Cruciform Specimen Test (Koyanagi et al., 2014) is probably the most promising test method for interface strength measurements involving a single-fiber composite specimen. This is because the method records the interface normal strength, which is slightly lower than the shear strength of the interface in general.

The fiber–matrix adhesion strength and the failure mechanism of the interface are also governed by the loading pattern, as revealed by both fiber pull-out and fiber push-out tests (Mäder et al., 2000). Figure 6.5 (Mäder et al., 2000) illustrates these test arrangements, which are self-explanatory. E-glass fiber and epoxy resin (DGEBA-based) composites are used for this test. The fibers include both untreated and sized fibers, with sizing accomplished by γ-APS (γ-aminopropyltriethoxysilane). The results were compared to record the effect of sizing. Figure 6.6 (Mäder et al., 2000) shows the results, F_d denoting the debond force. Both "stress-based" and "energy release rate" criteria are adopted for analysis of the results.

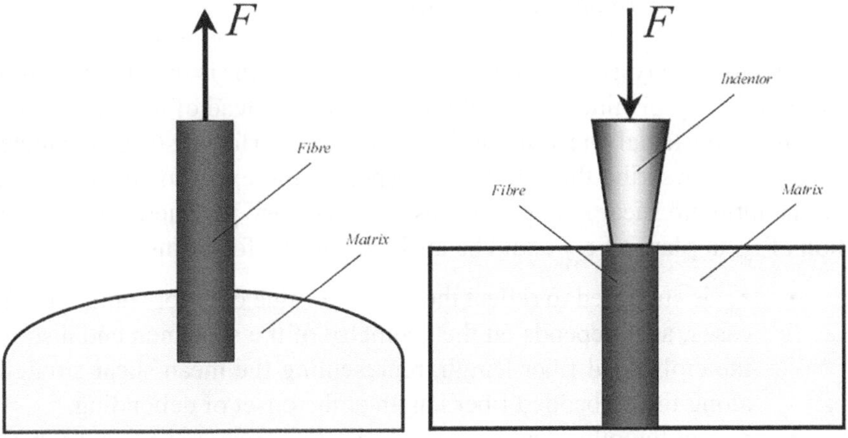

FIGURE 6.5 Schematic arrangements of the (a) pull-out and (b) push-out tests.

Source: Adapted from Mδder et al., 2000

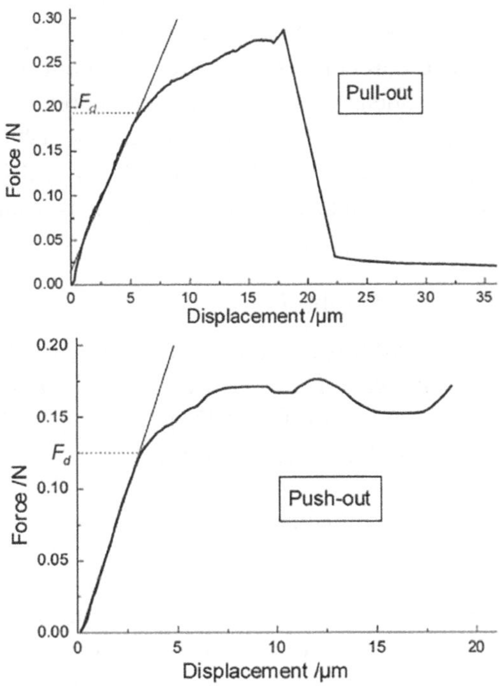

FIGURE 6.6 Typical force-displacement curves for the (a) pull-out and (b) push-out tests.

Source: Adapted from Mäder et al., 2000

6.3.1.1 *STRESS-BASED CRITERION*

Under this analysis criterion, the shear strength is calculated from experimentally obtained "F_d" values. However, instead of focusing on the apparent interfacial shear strength, τ_{app} as the interfacial shear parameter adequate to describe the interfacial response, the experimenters focused on the ultimate shear strength, τ_{ult}, as the necessary parameter. The selection of τ_{ult} in place of τ_{app} could be attributed to the following:

- τ_{app} is supposed to reflect the IFSS (τ) value only for some specific cases, as it depends on the geometry of the specimen and also on the embedded fiber length, representing the mean shear strength along the embedded fiber length at the onset of debonding.
- τ_{ult}, on the other hand, is the shear stress required to cause shear failure locally at a point of the interface and can be considered as the shear response of the interface in general

Thus, "τ_{ult}" characterizes the local interfacial shear strength near the crack tip. Gorbatkina (1992) first proposed the use of "τ_{ult}" which was later confirmed by many researchers based on the argument that "τ_{app}" typically yielded much smaller shear strength values compared to "τ_{ult}" (Desarmot and Favre, 1991; Ghiassi et al., 2014; Gorbatkina, 1992; Pisanova and Zhandarov, 1997; Pisanova et al., 1998; Zhandarov and Pisanova, 1997). τ_{ult}, as obtained from pull-out test results, was given as (Mäder et al., 2000):

$$\tau_{ult} = \frac{F_d \beta}{\pi d \tanh(\beta l_e)} + \tau_{therm} \tanh\left(\beta l_e/2\right) \tag{6.7}$$

where; "F_d" is the defined force obtained experimentally from the pull-out test; d is the fiber diameter; l_e is the fiber embedded length; β is the shear-lag parameter as defined by Nayfeh (1977); and τ_{therm} is the axial thermal stress given by Eqn. 6.8 (Mäder et al., 2000).

$$\tau_{therm} = E_f \beta d(\alpha_m - \alpha_f)\Delta T/4 \tag{6.8}$$

where; E_f is the tensile modules of the fiber; α_m is the coefficient of thermal expansion of matrix; α_f is the coefficient of thermal expansion of fiber; and T is the difference between the stress-free temperature and the test temperature.

The same Eqn. 6.7 holds good for the push-out test putting a minus sign before the τ_{therm} term to accommodate the direction changes of the axial thermal stress, considering "push-out" load direction as compared to that of the pull-out test. Also, the F_d value for push out test must be incorporated in the equation.

6.3.1.2 ENERGY-BASED CRITERION

On creation of a new surface due to fracture, energy is liberated. In the FRP composites the case can be considered parallel to the event of fiber-matrix debonding that creates the new-surface of separation between the fiber and the matrix. The energy release rate which is defined as the energy per unit debonded interfacial area or simply the critical energy release rate (G_{ic}) can also be used to characterize the fiber-matrix interface. For pull-out test, the "G_{ic}" is mathematically given as (Liu and Nairn, 1999):

$$G_{ic} = \frac{d}{4}\left\{C_{33s}\,\sigma_d^2 + 2D_{3s}\,\sigma_d\,\Delta T + \left[\frac{D_3^2}{C_{33}} + \frac{v_m\left(\alpha_f - \alpha_m\right)^2}{v_f\,A_o}\right]\Delta T^2\right\} \quad (6.9)$$

where; $\sigma_d = 4F_d(\pi d^2)$, is the stress applied to the fiber; v_f and v_m are volume fractions of the fiber and the matrix respectively, in the composite specimen; A_0, C_{33}, C_{33s}, D_3 and D_{3s} are constants, as defined by Liu and Nairn (1999).

Again, in order to express "G_{ic}" for the push-out test, the second term in Eqn. 6.9 has to have a minus sign, considering the directions of the externally applied stresses for pull-out and push-out tests. It may be appreciated that, as a relatively high τ_{ult} indicates better interfacial integrity, a relatively high G_{ic} value also indicates greater stability of the interface. This is because a higher G_{ic} indicates an increase in the stress intensity factor, resulting in an enhancement of the toughness of the composite material, which, obviously, would discourage any crack propagation.

The experimental results indicate that when the fibers used for composite fabrication are pre-treated with γ-APS, both push-out and pull-out bond strength measurements yield higher results as compared to the untreated fibers. However, both τ_{ult} and G_{ic} values are greater for the pull-out test compared to the push-out test. The higher τ_{ult} and G_{ic} values, which indicate better performance, integrity, and stability of the composite with γ-APS-treated fiber surfaces, may be attributed to improved γ-APS-treated fiber surface adhesion to the matrix (Pisanova and Mäder, 2000). Also, covalent bonds are formed between the epoxy groups of the matrix and the amino groups of γ-APS.

The higher values of τ_{ult} and G_{ic} in the pull-out tests as compared to the push-out tests are the consequences of the effect of loading pattern on the interfacial performances. In the pull-out test, the load is predominantly tensile in nature, in which there is a contraction of the fiber cross-sectional area. Here, the interface is sufficiently involved in stress transfer, and the failure is due to cohesive matrix failure at some distance from the fiber surface rather than brittle interfacial cracking. On the other hand, as seen from Figure 6.5(b) (Mäder et al., 2000), the indenter for loading in the push-out test is of a smaller diameter than that of the fiber. This generates a complex loading pattern involving compression, shear, and friction both inside the fiber and in the interphase surrounding it. Under these conditions, failure of the interface could be due to brittle crack growth, rendering relatively smaller values of the debonding forces.

The work to be accomplished to separate the fiber and the matrix, W_a defined as the work to be done to break unit area of any adhesion contact between the fiber and the matrix, also qualifies the interface. In terms of "τ_{ult}," W_a is given as W_a^* (Mäder et al., 2000) mathematically:

$$W_a^* = \sigma^{-1} \left(\frac{E_f}{E_m} \right)^{1/2} \tau_{ult} \tag{6.10}$$

where: σ is the reciprocal of the typical center-to-center distance involved in van der Waals and acid–base interactions, given approximately as equal to 2×10^9 m^{-1}.

$\left(\frac{E_f}{E_m} \right)^{1/2}$ is a normalizing factor to enable the comparison between all composite systems. It should be understood clearly, however, W_a, the reversible thermodynamic work of adhesion that qualifies the formation of adhesive contact and failure of the interface, is very rarely reversible. This is reflected in the fiber–matrix failure, which never generates the same surface qualities as the original surface before contact. It is of greater consequence for thermosetting polymer adhesion, wherein the cross-linking processes of polymerization are irreversible.

Use of Eqn. 6.10 (Mäder et al., 2000) yields results more realistic and relevant in defining the interface adhesion, as it characterizes the actual processes which could be irreversible, as witnessed in the case of actual adhesive joints.

6.4 RESIDUAL STRESS

Interfacial residual stresses are established in the FRP composites on account of the process-persuaded chemical reactions and shrinkages of the polymer matrix during polymerization on curing. It also finds a strong footing in the mismatching thermal expansion coefficients and elastic modulus of the constituent fiber and the matrix upon cooling from the cure temperature to room temperature, which brings about differential cooling contraction between the two. These chemical shrinkages and cooling contractions are also not unopposed (Chakraverty et al., 2014; Swain et al., 2019). Such volume contractions are strongly opposed and tend to be restricted by the accompanying reinforcement fibers. These opposing trends result in the generation of residual stresses in the interface as well as in the bulk matrix, leaving the FRP composite in a pre-load stress condition. In addition, the difference in the elastic modulus of the fiber and matrix may cause strain misfits between the two during loading, which could induce residual stresses in the FRP composites.

The process of curing consists of two distinct stages: gelation, pertaining to the transformation of the polymer's liquid phase into a rubbery state, and vitrification, the transition of the rubbery material into a glassy state, as shown in the T–T–T diagram in Figure 6.7 (Beura et al., 2018). Both of these transitions are associated with respective equilibrium elastic moduli, the rubbery modulus giving way to the glassy one. Therefore, curing not only results in chemical reactions leading to volume shrinkages due to polymerization, but also involves a change in the elastic modulus of the polymer. The combined effect of the two manifests as the development of shrinkage stresses (Swain et al., 2019).

On the other hand, the magnitude and nature of thermal cooling stresses depend on the constituents' contraction—of both the fiber and the matrix. The magnitude of the thermal contraction stresses depends on the intrinsic properties of the fiber and the matrix, which include their respective coefficients of thermal expansion (CTE), Poisson's ratios, and elastic moduli. During cooling from the curing temperature to room temperature, the matrix is restricted from undergoing full contraction by the fiber, which has a relatively small CTE and is bonded with it. Thus, the matrix is stressed in tension. Also, in addition to contracting on its own, the fiber embedded in the matrix is compressed by the matrix, which undergoes restricted contractions. This causes cooling contraction residual stresses

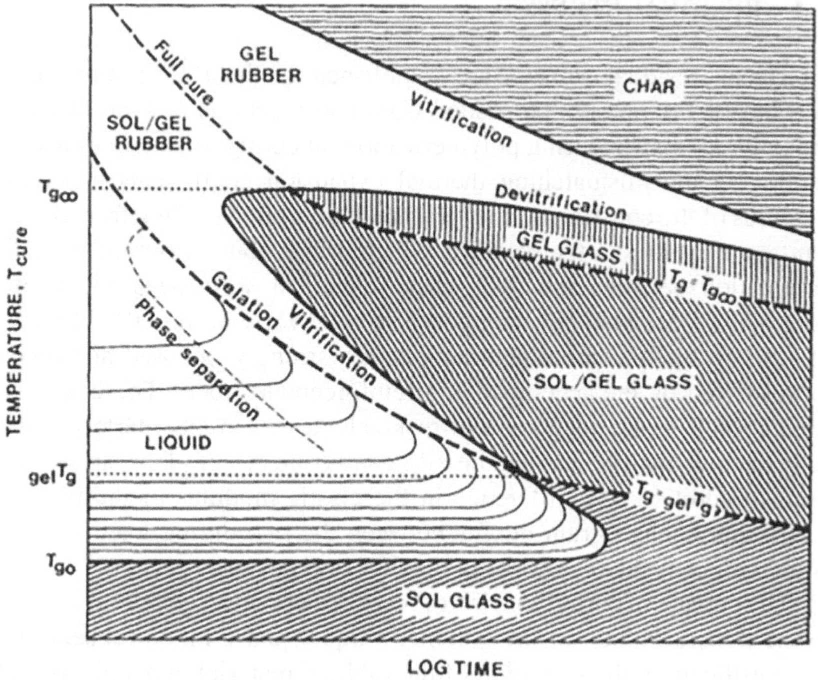

FIGURE 6.7 Time–temperature–transformation (T–T–T) isothermal cure diagram for a thermosetting resin system. The diagram depicts three critical temperatures T_{go}, gel T_g and T_{g1} and distinct states of matter.

Source: Adapted from Beura et al., 2018

6.4.1 *EFFECT OF RESIDUAL STRESS*

Depending on the nature and magnitude, interfacial stresses in FRP composites can either be detrimental or beneficial. However, the presence of residual stress is often detrimental to the stability of the composite. It is reported (Zhao et al., 2006) that tensile residual stresses in the matrix adversely affect the stability of the composite. It is probable that such residual stresses can assume high values and initiate matrix cracking and interfacial debonding, as depicted in Figure 6.8 (Zhao et al., 2006), prior to the loading of the composite. Mechanical property deterioration is a consequence of this. The stiffness and strength properties of the composite may be damaged irreparably. Furthermore, the microcracks can act as sources for macrocracks under the continuing effects of environmental agents such as humidity and thermal fluctuations. The average effects of residual stresses place the composite in a pre-load stressed condition and result in its premature failure, causing delamination, warpage, and cracking of the matrix, and so forth (Karataş and Gökkaya, 2018; Yang et al., 2013; Zhu and Lopez, 2014).

On the other hand, compressive interfacial residual stresses are claimed to be beneficial for a composite with low interfacial bond strength under transverse loading conditions (Mohamed et al., 2020; Zhang et al., 2016). In this case, the positive hoop stress around the cylindrical reinforcement fiber comes into play. The hoop stress enhances the pull-out energy losses, thereby increasing the pull-out stress requirements for extracting the fiber from the grips of the enveloping matrix, resulting in an augmentation of the toughness of the fiber–matrix composite.

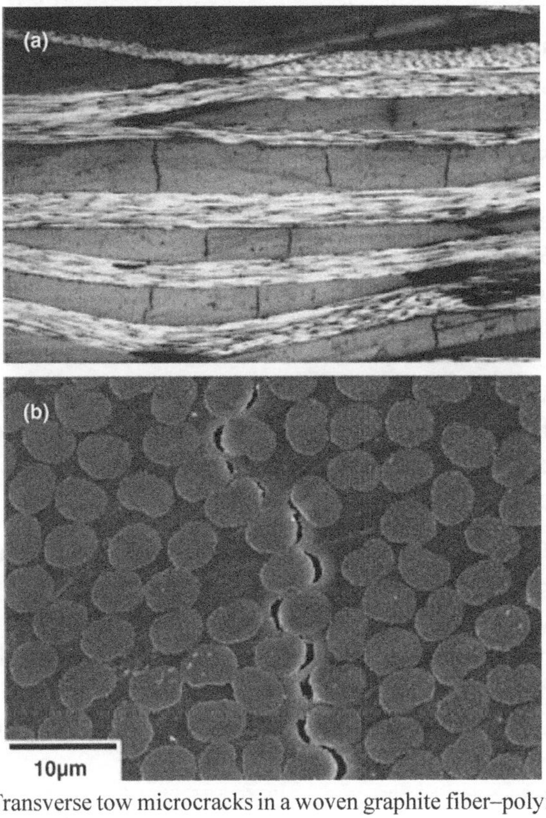

FIGURE 6.8 Transverse tow microcracks in a woven graphite fiber–polyimide composite prior to loading.

Source: Reprinted with permission from Zhao et al., 2006

Experimental findings by Zhao et al. (2006) predict a diminution and shifting of the "failure envelope" drawn for the FRP composite under investigation with the help of different biaxial strain ratios (strains in the two mutually perpendicular directions), with and without considering the interfacial residual stresses, as shown in Figure 6.9 (Zhao et al., 2006). The failure envelopes are constructed for two specific cases of high and low resin strength, with the compressive strength of the resin considered to be twice that of its tensile strength, as per listed provisions (Yang et al., 2020). Also, for the specific case, the stress/strain ratio is considered for 1% strain (shrinkage), and the fiber/matrix composite is considered to have been cured at 100 °C. The following is established from the plots (Figure 6.9):

- In compression, for both cases pertaining to high and low resin strengths, failure occurs earlier when residual stresses are considered, in comparison to the case when residual stresses are not considered. Thus, when loading is dominantly compressive, the presence of residual stresses adversely affects the stability of the FRP composite.
- In tension, however, as is clear from Figure 6.9 (Zhao et al., 2006), the composites with high and low resin strengths present two distinct situations, making the overall effect of the presence of residual stresses complicated.

  - **Case (a):** This pertains to the sample with low resin strength. Here, residual stresses lead to early failure and hence prove to be detrimental to the structural integrity of the FRP composite.

  - **Case (b):** This pertains to the composite sample with high resin strength. In this case, the shifting of the failure envelope when residual stresses are considered is seen to be clearly responsible for delayed failure. Therefore, residual stresses in high resin strength FRP composites may prove to be beneficial under tension-dominated loading, enhancing their structural reliability.

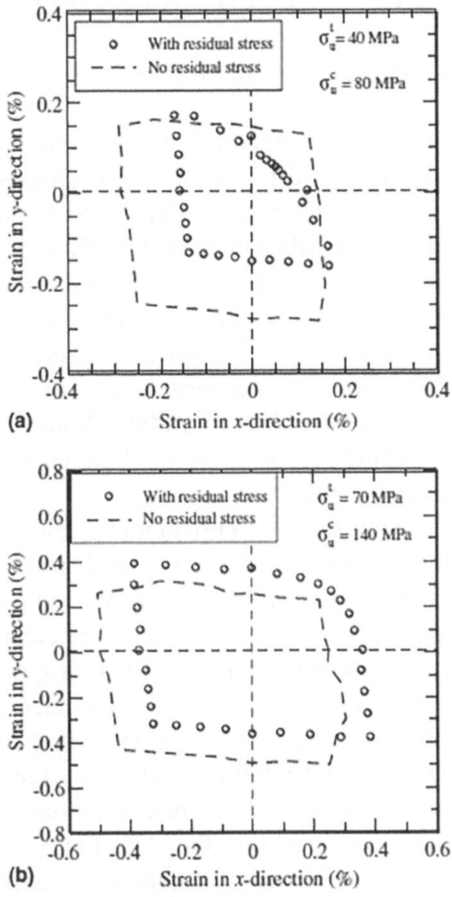

FIGURE 6.9 Failure envelopes for biaxial transverse normal loading: (a) $\sigma_u^c = 2\sigma_u^t = 80MPa$; (b) $\sigma_u^c = 2\sigma_u^t = 140MPa$.
Source: Reprinted with permission from Zhao et al., 2006

Zhao et al. (2007) report that the magnitude of residual stress in FRP composites depends on their fiber volume fraction and also on the pattern of fiber packing. The tendency of the fibers to oppose matrix contraction during process-induced cooling increases with increasing fiber content and contributes to the generation of higher levels of residual stresses. Also, a uniform fiber-packing pattern—such as in the case of a hexagonal array—reduces the overall fiber constraints and plays a role in decreasing the magnitude of residual stresses.

6.4.2 MEASUREMENT AND ANALYSIS OF STRESS

6.4.2.1 MEASUREMENT OF RESIDUAL STRESS

The direct measurement of residual stress at the fiber/matrix interface of FRP composites is difficult for two reasons. First, since the fiber diameters range from several microns to hundreds of microns, any measurement at the fiber–matrix interface requires extremely high resolution. Second, releasing the residual stress from the highly localized interface for direct measurement necessitates specialized procedures. On the basis of the above-mentioned limitations, several numerical and experimental methods used for analyzing residual stresses at fiber–matrix interfaces (Ju, 2000; Warrier et al., 1999; Zou et al., 1998) do not fully resolve the challenges of direct interfacial residual stress measurement.

Considering these constraints, a novel method was proposed by Xing et al. (2004) for determining interfacial residual stresses in FRP composites. The method addresses both the high-resolution requirement and the difficulty of residual stress release. It employs the electron moiré method (Dally and Read, 1993; Read and Dally, 1994; Xie et al., 2000), which involves fabrication of a grid on the specimen surface using scanning electron microscopy (SEM) and an electron beam. The grid has a high frequency—on the order of 10,000 lines/mm—on the specimen surface. Such high resolution enables the direct measurement of strain (and hence stress) in the focused, localized interfacial region.

Residual stress release at the interface is actuated using the fiber push-out technique (Chandra and Ananth, 1995; Chandra and Ghonem, 2001; Galbraith et al., 1996). As illustrated in Figure 6.10 (Xing et al., 2004), compressive stresses at the top face of the fiber—applied by an indenter—initiate interfacial shearing until the entire interface fails and the fiber is completely pushed out. The residual stress at the interface is thereby released, allowing for measurement of the resulting strain using the high-resolution electron moiré method. Typically, residual strain (and stress) is measured in the polymer matrix, since it has a lower modulus than the fiber and thus undergoes greater deformation. The residual strain thus measured can be converted to residual stress using Hooke's law, provided the stress is below the material's yield point.

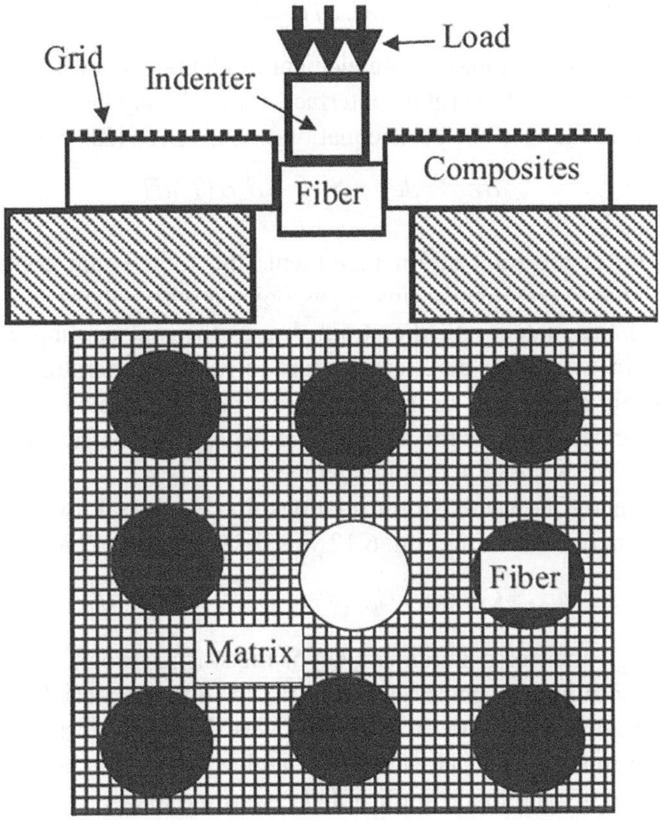

FIGURE 6.10 Illustration of the process for releasing residual stress.

Mathematically, the residual stress is given as (Xing et al., 2004):

$$\sigma_{ij} = \frac{E}{(1+\upsilon)(1-2\upsilon)}[(1-\upsilon)\varepsilon_{ij} + 2\upsilon\varepsilon_{ji}] \qquad (6.11)$$

where; σ_{ij} and σ_{ji} are the longitudinal and the transverse stresses; E is the Young's modulus; v is the Poisson's Ratio; and ε_{ij} and ε_{ji} denotes residual strain components in the longitudinal and the transverse directions.

The method has been successfully used by Xing et al. (2004) to measure residual stresses at SiC/Ti-15-3 composite interfaces and in Boron/Epoxy laminates. Thus, the method is deemed suitable for interfacial stress measurements in both metal-matrix and polymer-matrix composites.

6.4.2.2 ANALYSIS OF RESIDUAL STRESS

Zhao et al. (2006) adopted finite element modeling to develop basic equations for the stress (strain) at the interface, for analyzing the stresses in terms of the related factors. The basic equation was given as (Zhao et al., 2006):

$$d\varepsilon_{ij} = de_{ij} + \delta_{ij} ds + \delta_{ij}\alpha(T)dT \tag{6.12}$$

where; $d\varepsilon_{ij}$ is the total strain increment (the total strain comprises the strain related to chemical shrinkage associated with polymerization cross-linking of the matrix and the strain due to thermal cooling contraction of the matrix from the cure temperature to room temperature); de_{ij} is the elastic strain increment; ds is the chemical reaction related free shrinkage strain increment; $\alpha(T)$ is the temperature dependent coefficient of thermal expansion; dT is the temperature change.

Also, using the stress-strain relationship (Hooke's Law) the equation for stress increment using Eqn. 6.12 was developed as (Zhao et al., 2006):

$$d\sigma_{ij} = C_{ijkl}\, de_{kl} = C_{ijkl}\left\{d\varepsilon_{ij} - \delta_{ij}\, ds - \sigma_{ij}\, \alpha(T)\, dT\right\} \tag{6.13}$$

where; $d\sigma_{ij}$ is the stress increment; C_{ijkl} is the stiffness coefficient related to Young's modulus "E" and Poisson's ratio "υ" of the material.

This Eqn. 6.13 (Zhao et al., 2006) was developed primarily for the polymeric matrix resin. However, by neglecting the chemical shrinkage strain, it can also be made applicable to isotropic fibers.

These equations were used to estimate and analyze interfacial residual stresses in FRP composites. For this purpose, in the first step, the shrinkage residual stress for the resin was calculated for a given amount of resin shrinkage. The resin shrinkage was considered to lie within the range investigated by several experimenters (Beura et al., 2020; Li et al., 2004; Sakaguchi et al., 2004; Singer et al., 2019). The second step involved the estimation and analysis of thermal cooling contraction stress (Amir-Ahmadi and Ghasemi, 2020; Mohamed et al., 2020; Salling et al., 2020).

Finite element analysis was conducted for the shrinkage stress calculations by incrementally applying the total shrinkage strain using Eqns. 6.12 and 6.13 (Zhao et al., 2006). Similarly, residual stress calculations pertaining to thermal cooling contraction were conducted by setting the shrinkage strain to zero, in which case only the thermal cooling contractions were considered for the calculations, again using the two basic Eqns. 6.12 and 6.13 (Zhao et al., 2006).

For analytical purposes in failure prediction, both the matrix and the fiber were considered to be isotropic, for which the maximum principal stress theory is applicable. The maximum failure stress criterion was given as (Zhao et al., 2006):

$$\sigma_{\max} \geq \sigma_4^t \text{ and } \sigma_{\min} \geq \sigma_4^c \tag{6.14}$$

where; σ_{max} is the maximum principal stress; σ_{min} is the minimum principal stress; σ_4^t is the tensile strength of the material; and σ_4^c is the compressive strength of the material.

Obviously, the material fails when the yield criteria are satisfied, marking the rupture of the unit cell considered. To represent the contributions of both chemical shrinkage and thermal cooling contraction toward the maximum principal residual stress along the interface, a specific case with 1% shrinkage strain and a curing temperature of 100°C was considered. The distribution of the maximum principal shear stress of the resin was plotted, as presented in Figure 6.11 (Zhao et al., 2006).

The contribution of chemical shrinkage stresses was seen to be much less than that of thermal cooling contraction stresses in building up the maximum principal residual stress of the resin. This might be attributed to the relatively low curing temperature (100°C) of the resin (Li et al., 2019b; Tsang and Taylor, 2019), which corresponds to lower values of the modulus.

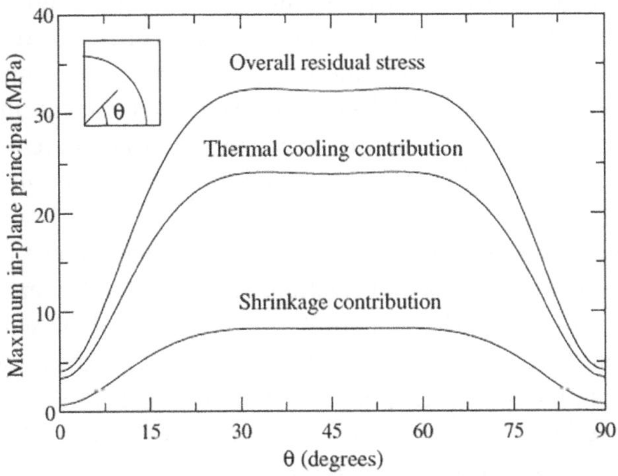

FIGURE 6.11 Distribution of the maximum principal residual stress in the resin along the fiber/matrix interface for 1% shrinkage strain and 100°C curing temperature.

Source: Reprinted with permission from Zhao et al., 2006

6.5 CONCLUDING REMARKS

A holistic, quality-driven utilization of FRP composites requires a thorough analysis of their intrinsic characteristics. Needless to say, the fiber–matrix interface in the FRP composite governs stress distribution and sharing between the constituent phases. Besides the inherent properties of the reinforcement and matrix phases, it is the characteristics of the interface that determine the mechanical performance of the FRP composite. Therefore, the interface in FRP composites deserves special attention when exploring the possibilities of engineering the material for specific end use.

The FRP composite performs in the way the interface performs, with interfacial bond strength being a direct measure of the structural integrity of the FRP composite material. A strong interfacial bond imparts good toughness, shear strength, and compression strength to the composite. On the other hand, a weak interface—potentially with the presence of voids—may provide a pathway for moisture ingression into the composite. The voids at the interface, in addition to assisting moisture wicking, may also contribute to stress concentration in the material, which can lead to a loss of stability. However, the interfacial bond strength must be optimized; both excessively high and low values can be detrimental to the structural health of the composite.

Both the work of adhesion, "W_a" and the interfacial shear stress (IFSS), τ, are important tools for assessing interfacial bond strength, which can be mathematically expressed in terms of several operational parameters. It is also important to note that the loading pattern greatly influences the interfacial bond strength in FRP composites. In this context, the contribution of energy released per unit of debonded interfacial area, G_{ic}, cannot be ignored. For strong interfacial bonding "G_{ic}" should assume relatively high values.

Wettability concerns at the fiber–matrix interface constitute another important aspect of interface evaluation. Good wettability ensures strong fiber–matrix adhesion. The wettability parameters include work of adhesion, fiber surface energy, and interfacial energy. Analysis and control of these parameters—by adopting methods to enhance the fiber surface energy and reduce the fiber–matrix contact angle, θ, for better wetting—pave the way for tailoring the interface.

The role of residual stresses at the interface has also been analyzed in the present article. It is of great concern that FRP composites may exist in a pre-load stressed condition due to residual stresses at the interface, which develop as a result of the two-stage curing operations during matrix polymerization. However, under certain favorable conditions, the presence of residual stresses may actually enhance interfacial bond strength. It may also be noted that changes in elastic modulus during curing can lead to strain misfits, which adversely affect the performance of the composite under load.

KEYWORDS

- **compatibility**
- **heterogeneous material**
- **Hooke's law**
- **interfacial bond**
- **Poisson's ratio**
- **polymer matrix**
- **residual stress release**
- **stress accumulation**
- **stress-strain relationship**
- **wettability**

REFERENCES

Amir-Ahmadi, S., & Ghasemi, A. R. (2020). Experimental investigation of cooling conditions, MWCNTs and mandrel diameter effects on the thermal residual stresses of multi-layered filament-wound composite pipes. *Journal of Composite Materials*, *54*(30), 4773–4786.

Barczewski, M., Matykiewicz, D., Andrzejewski, J., & Skorczewska, K. (2016). Application of waste bulk moulded composite (BMC) as a filler for isotactic polypropylene composites. *Journal of Advanced Research*, *7*(3), 373–380.

Beura, S., Chakraverty, A. P., Sarangi, S., Mohanty, U. K., & Thatoi, D. N. (2020). Characterization of Plasma Cured E-Glass-Epoxy composite. *Materials Today: Proceedings*, *22*, 3082–3088.

Beura, S., Chakraverty, A. P., Thatoi, D. N., Mohanty, U. K., & Mohapatra, M. (2021a). Failure modes in GFRP composites assessed with the aid of SEM fractographs. *Materials Today: Proceedings*, *41*, 172–179.

Beura, S., Sahoo, S. R., Thatoi, D. N., Mohanty, U. K., & Chakraverty, A. P. (2021b). Stability of GFRP composites with varied fractions of reinforcement exposed to ageing processes outdoors. *Polymers and Polymer Composites*, *29*(9), 1485–1494.

Beura, S., Thatoi, D. N., Chakraverty, A. P., & Mohanty, U. K. (2018). Impact of the ambiance on GFRP composites and role of some inherent factors: A review report. *Journal of Reinforced Plastics and Composites*, *37*(8), 533–547.

Carloni, C., Santandrea, M., & Imohamed, I. A. O. (2017). Determination of the interfacial properties of SRP strips bonded to concrete and comparison between single-lap and notched beam tests. *Engineering Fracture Mechanics*, *186*, 80–104.

Chakraverty, A. P., Beura, S., Mohanty, U. K., Mishra, S. C., & Biswal, B. B. (2020a, March). Gamma-Irradiation of E-glass/epoxy composite: a study of its mechanical and thermal sustainability. In *Materials Science Forum* (Vol. 978, pp. 296–303). Trans Tech Publications Ltd.

Chakraverty, A. P., Dash, S., Maharana, H. S., Beura, S., & Mohanty, U. K. (2020b). A novel investigation on durability of GRE composite pipe for prolonged sea water transportation. *Composites Communications*, *17*, 42–50.

Chakraverty, A. P., Mohanty, U. K., Mishra, S. C., & Biswal, B. B. (2017, February). Effect of hydrothermal immersion and hygrothermal conditioning on mechanical properties of GRE composite. In *IOP Conference Series: Materials Science and Engineering* (Vol. 178, No. 1, p. 012013). IOP Publishing.

Chakraverty, A. P., Mohanty, U. K., Mishra, S. C., & Biswal, B. B. (2014). Evaluation of Glass fiber Reinforced Polymer Composites Under the Exposure to Up and Down-Thermal Shocks. *Advanced Science Letters*, *20*(3–4), 676–679.

Chakraverty, A. P., Parida, D., Dash, S., Parida, M., Beura, S., & Mohanty, U. K. (2020c). Thermo-mechanical response of post-plasma irradiated E-glass fiber/epoxy composite. *Materials Today: Proceedings*, *33*, 4956–4960.

Chandra, N., & Ananth, C. R. (1995). Analysis of interfacial behavior in MMCs and IMCs by the use of thin-slice push-out tests. *Composites Science and Technology*, *54*(1), 87–100.

Chandra, N., & Ghonem, H. (2001). Interfacial mechanics of push-out tests: theory and experiments. *Composites part A: applied science and manufacturing*, *32*(3–4), 575–584.

Chen, J., & Chouw, N. (2018). Effect of the interface condition on the bond between flax FRP tube and coconut fiber reinforced concrete composites. *Construction and Building Materials*, *167*, 597–604.

Chen, P., Lu, C., Yu, Q., Gao, Y., Li, J., & Li, X. (2006). Influence of fiber wettability on the interfacial adhesion of continuous fiber-reinforced PPESK composite. *Journal of Applied Polymer Science*, *102*(3), 2544–2551.

Dally, J. W., & Read, D. T. (1993). Electron beam moiré. *Experimental Mechanics*, *33*, 270–277.

Desarmot, G., & Favre, J. P. (1991). Advances in pull-out testing and data analysis. *Composites Science and Technology*, *42*(1–3), 151–187.

Egoshi, T., Ishikawa, T., Tezura, M., & Kizuka, T. (2020). Transmission electron microscopy of unidirectional carbon fiber reinforced plastics at on-axis tension. *Materialia*, *14*, 100907.

Fitriah, S. N., Majid, M. A., Ridzuan, M. J. M., Daud, R., Gibson, A. G., & Assaleh, T. A. (2017). Influence of hydrothermal ageing on the compressive behavior of glass fiber/epoxy composite pipes. *Composite Structures*, *159*, 350–360.

Fuentes, C. A., Brughmans, G., Tran, L. Q. N., Dupont-Gillain, C., Verpoest, I., & Van Vuure, A. W. (2015). Mechanical behavior and practical adhesion at a bamboo composite interface: Physical adhesion and mechanical interlocking. *Composites Science and Technology*, *109*, 40–47.

Galbraith, J. M., Rhyne, E. P., Koss, D. A., & Hellmann, J. R. (1996). The interfacial failure sequence during fiber pushout in metal matrix composites. *Scripta materialia*, *35*(4), 543–549.

Ghiassi, B., Verstrynge, E., Lourenço, P. B., & Oliveira, D. V. (2014). Characterization of debonding in FRP-strengthened masonry using the acoustic emission technique. *Engineering Structures*, *66*, 24–34.

Goh, G. D., Dikshit, V., Nagalingam, A. P., Goh, G. L., Agarwala, S., Sing, S. L., Wei, J., & Yeong, W. Y. (2018). Characterization of mechanical properties and fracture mode of additively manufactured carbon fiber and glass fiber reinforced thermoplastics. *Materials & Design*, *137*, 79–89.

Gorbatkina, Y. A. (1992). Adhesive strength in fibre-polymer systems. *(No Title)*.

Harding, P. H., & Berg, J. C. (1997). The role of adhesion in the mechanical properties of filled polymer composites. *Journal of Adhesion Science and Technology*, *11*(4), 471–493.

Hoecker, F., & Karger-Kocsis, J. (1996). Surface energetics of carbon fibers and its effects on the mechanical performance of CF/EP composites. *Journal of Applied Polymer Science*, *59*(1), 139–153.

Hull, D. (1998). An Introduction to Composite Materials.

Ismael, M. K., Abdulrazaq, E. A., Hassoni, S. M., & Haleem, A H. (2020). Shear strength of steel fiber/polyepoxide resin interface using single fiber pull-out tests. *Reinforced Plastics*, *64*(2), 103–108.

Jacobasch, H. J., Grundke, K., Mäder, E., Freitag, K. H., & Panzer, U. (1992). Application of the surface free energy concept in polymer processing. *Journal of adhesion science and technology*, *6*(12), 1381–1396.

Jacobasch, H. J., Grundke, K., Uhlmann, P., Simon, F., & Mäder, E. (1995). Comparison of surface-chemical methods for characterizing carbon fiber-epoxy resin composites. *Composite Interfaces*, *3*(4), 293–320.

Jeyakumar, R., Sampath, P. S., Ramamoorthi, R., & Ramakrishnan, T. (2017). Structural, morphological and mechanical behavior of glass fiber reinforced epoxy nanoclay composites. *The International Journal of Advanced Manufacturing Technology*, *93*, 527–535.

Ju, D. Y. (2000). Simulation of thermo-mechanical behavior and interfacial stress of metal matrix composite under thermal shock process. *Composite Structures*, *48*(1–3), 113–118.

Karataş, M. A., & Gökkaya, H. (2018). A review on machinability of carbon fiber reinforced polymer (CFRP) and glass fiber reinforced polymer (GFRP) composite materials. *Defence Technology*, *14*(4), 318–326.

Koyanagi, J., Ogihara, S., Nakatani, H., Okabe, T., & Yoneyama, S. (2014). Mechanical properties of fiber/matrix interface in polymer matrix composites. *Advanced Composite Materials*, *23*(5–6), 551–570.

Kumosa, L., Benedikt, B., Armentrout, D., & Kumosa, M. (2004). Moisture absorption properties of unidirectional glass/polymer composites used in composite (non-ceramic) insulators. *Composites Part A: Applied Science and Manufacturing, 35*(9), 1049–1063.

Kwon, D. J., Kim, J. H., Park, S. M., Kwon, I. J., DeVries, K. L., & Park, J. M. (2019). Damage sensing, mechanical and interfacial properties of resins suitable for new CFRP rope for elevator applications. *Composites Part B: Engineering, 157*, 259–265.

Lee, Y. S., & Lee, B. K. (2002). Surface properties of oxyfluorinated PAN-based carbon fibers. *Carbon, 40*(13), 2461–2468.

Li, C., Potter, K., Wisnom, M. R., & Stringer, G. (2004). In-situ measurement of chemical shrinkage of MY750 epoxy resin by a novel gravimetric method. *Composites Science and Technology, 64*(1), 55–64.

Li, H., Chen, G., Su, H., Li, D., Sun, L., & Yang, J. (2019a). Effect of the stoichiometric ratio on the crosslinked network structure and cryogenic properties of epoxy resins cured at low temperature. *European Polymer Journal, 112*, 792–798.

Li, X., Tao, R., Alfano, M., & Lubineau, G. (2020). How variability in interfacial properties results in tougher bonded composite joints by triggering bridging. *International Journal of Solids and Structures, 191*, 87–98.

Li, Y., Li, W., Shao, J., Deng, Y., Kou, H., Ma, J., Zhang, X., Zhang, X., Chen, L., & Qu, Z. (2019b). Modeling the effects of interfacial properties on the temperature dependent tensile strength of fiber reinforced polymer composites. *Composites Science and Technology, 172*, 74–80.

Liu, C. H., & Nairn, J. A. (1999). Analytical and experimental methods for a fracture mechanics interpretation of the microbond test including the effects of friction and thermal stresses. *International journal of adhesion and adhesives, 19*(1), 59–70.

Liu, H., Liu, J., Ding, Y., Zhou, J., Kong, X., Blackman, B. R., Kinloch, A. J., Falzon, B. G., & Dear, J. P. (2020). Effects of impactor geometry on the low-velocity impact behavior of fiber-reinforced composites: an experimental and theoretical investigation. *Applied Composite Materials, 27*, 533–553.

Liu, L., Jia, C., He, J., Zhao, F., Fan, D., Xing, L., Wang, M., Wang, F., Jiang, Z., & Huang, Y. (2015). Interfacial characterization, control and modification of carbon fiber reinforced polymer composites. *Composites Science and Technology, 121*, 56–72.

Lu, C., Wang, J., Lu, X., Zheng, T., Liu, Y., Wang, X., Zhang, D., & Seveno, D. (2019). Wettability and interfacial properties of carbon fiber and poly (ether ether ketone) fiber hybrid composite. *ACS Applied Materials & Interfaces, 11*(34), 31520-31531.

Luner, P. E., & Oh, E. (2001). Characterization of the surface free energy of cellulose ether films. *Colloids and Surfaces A: Physicochemical and Engineering Aspects, 181*(1–3), 31–48.

Mäder, E., Grundke, K., Jacobasch, H. J., & Wachinger, G. (1994). Surface, interphase and composite property relations in fiber-reinforced polymers. *Composites, 25*(7), 739–744.

Mittal, K. L. (Ed.). (2014). *Adhesion Measurement of Films and Coatings*. CRC Press.

Mohamed, M., Brahma, S., Ning, H., & Pillay, S. (2020). Development of beneficial residual stresses in glass fiber epoxy composites through fiber prestressing. *Journal of Reinforced Plastics and Composites, 39*(13–14), 487–498.

Mδder, E., Zhou, X. F., Pisanova, E., Zhandarov, S., & Nutt, S. R. (2000). Characterization of the interfacial bond strength between glass fiber and epoxy resin using the pull-out and push-out techniques. *Advanced Composites Letters, 9*(3), 096369350000900304.

Nardin, M., & Schultz, J. (1993). Relationship between fiber-matrix adhesion and the interfacial shear strength in polymer-based composites. *Composite Interfaces*, *1*(2), 177–192.

Nardin, M., Asloun, E. M., & Schultz, J. (1990). Physico-chemical interactions between carbon fibers and PEEK. In *Controlled Interphases in Composite Materials: Proceedings of the Third International Conference on Composite Interfaces (ICCI-III) held on May 21–24, 1990 in Cleveland, Ohio, USA* (pp. 285–293). Springer Netherlands.

Nardin, M., Asloun, E. M., & Schultz, J. (1991). Study of the carbon fiber–poly (ether–ether–ketone)(PEEK) interfaces, 2: relationship between interfacial shear strength and adhesion energy. *Polymers for Advanced Technologies*, *2*(3), 115–122.

Nayfeh, A. H. (1977). Thermomechanically induced interfacial stresses in fibrous composites. fiber *Science and Technology*, *10*(3), 195–209.

Packham, D. E. (1996). Work of adhesion: contact angles and contact mechanics. *International Journal of Adhesion and Adhesives*, *16*(2), 121–128.

Park, J. M., Kim, D. S., & Kim, S. R. (2003). Improvement of interfacial adhesion and nondestructive damage evaluation for plasma-treated PBO and Kevlar fibers/epoxy composites using micromechanical techniques and surface wettability. *Journal of Colloid and Interface Science*, *264*(2), 431–445.

Park, J. M., Kim, D. S., Kong, J. W., Kim, M., Kim, W., & Park, I. S. (2002). Interfacial adhesion and microfailure modes of electrodeposited carbon fiber/epoxy–PEI composites by microdroplet and surface wettability tests. *Journal of Colloid and Interface Science*, *249*(1), 62–77.

Parveen, S., Pichandi, S., Goswami, P., & Rana, S. (2020). Novel glass fiber reinforced hierarchical composites with improved interfacial, mechanical and dynamic mechanical properties developed using cellulose microcrystals. *Materials & Design*, *188*, 108448.

Pisanova, E. V., & Zhandarov, S. F. (1997). On the mechanism of failure in microcomposites consisting of single glass fibers in a thermoplastic matrix. *Composites Science and Technology*, *57*(8), 937–943.

Pisanova, E., & Mäder, E. (2000). Acid–base interactions and covalent bonding at a fiber–matrix interface: contribution to the work of adhesion and measured adhesion strength. *Journal of Adhesion Science and Technology*, *14*(3), 415–436.

Pisanova, E., Dutschk, V., & Lauke, B. (1998). Work of adhesion and local bond strength in glass fiber-thermoplastic polymer systems. *Journal of Adhesion Science and Technology*, *12*(3), 305–322.

Prusty, R. K., Rathore, D. K., & Ray, B. C. (2017). CNT/polymer interface in polymeric composites and its sensitivity study at different environments. *Advances in Colloid and Interface Science*, *240*, 77–106.

Qasem, A., Sallam, Y. S., Eldien, H. H., & Ahangarn, B. H. (2020). Bond-slip behavior between ultra-high-performance concrete and carbon fiber reinforced polymer bars using a pull-out test and numerical modelling. *Construction and Building Materials*, *260*, 119857.

Rafiee, R., & Mazhari, B. (2016). Simulation of the long-term hydrostatic tests on glass fiber reinforced plastic pipes. *Composite Structures*, *136*, 56–63.

Read, D. T., & Dally, J. W. (1994). Electron beam moiré study of fracture of a glass fiber reinforced plastic composite, 402–409.

Saïhi, D., El-Achari, A., Ghenaim, A., & Cazé, C. (2002). Wettability of grafted poly (ethylene terephthalate) fibers. *Polymer testing*, *21*(6), 615–618.

Sakaguchi, R. L., Wiltbank, B. D., & Murchison, C. F. (2004). Prediction of composite elastic modulus and polymerization shrinkage by computational micromechanics. *Dental Materials*, *20*(4), 397–401.

Salling, F. B., Sandberg, M., Spangenberg, J., & Hattel, J. H. (2020, October). Numerical and experimental analyses in composites processing: impregnation, heat transfer, resin cure and residual stresses. In *IOP Conference Series: Materials Science and Engineering* (Vol. 942, No. 1, p. 012003). IOP Publishing.

Sheu, G. S., & Shyu, S. S. (1994). Surface modification of Kevlar 149 fibers by gas plasma treatment. Part II. Improved interfacial adhesion to epoxy resin. *Journal of Adhesion Science and Technology*, *8*(9), 1027–1042.

Shi, L., Song, G., Li, P., Li, X., Pan, D., Huang, Y., Ma, L., & Guo, Z. (2021). Enhancing interfacial performance of epoxy resin composites via in-situ nucleophilic addition polymerization modification of carbon fibers with hyperbranched polyimidazole. *Composites Science and Technology*, *201*,108522.

Shin, P. S., Kwon, D. J., Kim, J. H., Lee, S. I., DeVries, K. L., & Park, J. M. (2017). Interfacial properties and water resistance of epoxy and CNT-epoxy adhesives on GFRP composites. *Composites Science and Technology*, *142*, 98–106.

Singer, G., Sinn, G., Lichtenegger, H. C., Veigel, S., Zecchini, M., & Wan-Wendner, R. (2019). Evaluation of in-situ shrinkage and expansion properties of polymer composite materials for adhesive anchor systems by a novel approach based on digital image correlation. *Polymer Testing*, *79*, 106035.

Swain, S., Beura, S., Thatoi, D. N., Chakraverty, A. P., & Mohanty, U. K. (2019). Durability of GFRP composite exposed to outdoors weathering. *Composites Communications*, *13*, 22–29.

Swolfs, Y., Verpoest, I., & Gorbatikh, L. (2016). A review of input data and modelling assumptions in longitudinal strength models for unidirectional fiber-reinforced composites. *Composite Structures*, *150*, 153–172.

Tan, W., Naya, F., Yang, L., Chang, T., Falzon, B. G., Zhan, L., Molina-Aldareguía, J. M., González, C., & Llorca, J. (2018). The role of interfacial properties on the intralaminar and interlaminar damage behavior of unidirectional composite laminates: Experimental characterization and multiscale modelling. *Composites Part B: Engineering*, *138*, 206–221.

Tsang, W. L., & Taylor, A. C. (2019). Fracture and toughening mechanisms of silica-and core–shell rubber-toughened epoxy at ambient and low temperature. *Journal of Materials Science*, *54*(22), 13938–13958.

Ulkem, I., & Schreiber, H. P. (1994). The role of interactions at interfaces of glass-fiber reinforced composites. *Composite Interfaces*, *2*(4), 253–263.

Warrier, S. G., Rangaswamy, P., Bourke, M. A. M., & Krishnamurthy, S. (1999). Assessment of the fiber/matrix interface bond strength in SiC/Ti-6Al-4V composites. *Materials Science and Engineering: A*, *259*(2), 220–227.

Wu, G. M. (2004). Oxygen plasma treatment of high performance fibers for composites. *Materials Chemistry and Physics*, *85*(1), 81–87.

Wu, S. (2017). *Polymer Interface and Adhesion*. Routledge.

Xie, H., Kishimoto, S., & Shinya, N. (2000). Fabrication of high-frequency electron beam Moire grating using multi-deposited layer techniques. *Optics & Laser Technology, 32*(5), 361–367.

Xing, Y. M., Kishimoto, S., Tanaka, Y., & Shinya, N. (2004). A novel method for determining interfacial residual stress in fiberreinforced composites. *Journal of Composite Materials, 38*(2), 137–148.

Yang, H., Lei, H., Lu, G., Zhang, Z., Li, X., & Liu, Y. (2020). Energy absorption and failure pattern of hybrid composite tubes under quasi-static axial compression. *Composites Part B: Engineering, 198*, 108217.

Yang, L., Yan, Y., Ma, J., & Liu, B. (2013). Effects of inter-fiber spacing and thermal residual stress on transverse failure of fiber-reinforced polymer–matrix composites. *Computational Materials Science, 68*, 255–262.

Yoo, D. Y., Kim, S., Park, G. J., Park, J. J., & Kim, S. W. (2017). Effects of fiber shape, aspect ratio, and volume fraction on flexural behavior of ultra-high-performance fiber-reinforced cement composites. *Composite Structures, 174*, 375–388.

Zhandarov, S. F., & Pisanova, E. V. (1997). The local bond strength and its determination by fragmentation and pull-out tests. *Composites Science and Technology, 57*(8), 957–964.

Zhang, Y., Xu, F., Zhang, C., Wang, J., Jia, Z., Hui, D., & Qiu, Y. (2016). Tensile and interfacial properties of polyacrylonitrile-based carbon fiber after different cryogenic treated condition. *Composites Part B: Engineering, 99*, 358–365.

Zhao, L. G., Warrior, N. A., & Long, A. C. (2006). A micromechanical study of residual stress and its effect on transverse failure in polymer–matrix composites. *International Journal of Solids and Structures, 43*(18–19), 5449–5467.

Zhao, L., Li, Y., Zhang, J., Zhou, L., & Hu, N. (2018). A novel material degradation model for unidirectional CFRP composites. *Composites Part B: Engineering, 135*, 84–94.

Zhao, L., Warrior, N. A., & Long, A. C. (2007, July). Analysis of residual stress in fiber-reinforced polymer composites. In *16th International Conference on Composite Materials*.

Zhou, A., Qin, R., Feo, L., Penna, R., & Lau, D. (2017). Investigation on interfacial defect criticality of FRP-bonded concrete beams. *Composites Part B: Engineering, 113*, 80–90.

Zhu, J., & Lopez, M. M. (2014). Performance of a lightweight GFRP composite bridge deck in positive and negative bending regions. *Composite Structures, 113*, 108–117.

Zou, H., Liu, J., Ding, D. H., Wang, R., Froyen, L., & Delaey, L. (1998). Determination of interfacial residual stress field in an Al–Al_2O_3 composite using convergent-beam electron diffraction technique. *Ultramicroscopy, 72*(1–2), 1–15.

CHAPTER 7

Mechanical and Corrosion Behavior of Friction Stir Processed AA7075 Hybrid Surface Composite

RAMESH NAIDU,[1] SRINIVASU ROKKALA,[2] VIJAY KUMAR AMBATI,[2] AMIT KUMAR MEHAR,[2,*] and D. S. CHAITANYA KISHORE[1]

[1]*Department of Mechanical Engineering, JNTU Anantapur, Anantapur, Andhra Pradesh, India*

[2]*Department of Mechanical Engineering, Raghu Engineering College (A), Dakamarri, Visakhapatnam, Andhra Pradesh, India*

*Corresponding author

ABSTRACT

Surface composites have been crucial to the fabrication of structures with applications in aeronautical industries and also in defense sectors. Through strong, localized plastic deformation, friction stir processing (FSP) can alter a metal's characteristics. In an ideal world, this technique produces a microstructure with tiny, equiaxed grains and mixes the material without altering the phase (either by melting or another method). The metal's fatigue and tensile strengths are also improved by FSP. A hybrid nano-composite (NC) can be fashioned by dispersing inorganic nanoparticles into a metal matrix. In this research work, a surface composite based on AA7075 alloy with zinc as a primary alloying element is fabricated using FSP. Among the base material's main drawbacks are its attributes like low melting point, low strength, and inadequate wear behavior. The fabricated composite was found to have improved properties with respect to hardness and corrosion, along with reduced strength. In this present study, we are using tool rotation speed and feed factors, percentage of BN particles, and carbon nanotube (CNT) as factors to evaluate the properties of the surface

Advances in Industrial Manufacturing and Materials. Integrating a New Era of Computational Modeling and Machine Learning. Debesh Mishra and Swagatika Mishra (Eds.)

DOI: 10.1201/9781779640475-7

composite, using SEM, optical and corrosion tests, and tensile tests to measure the same.

7.1 INTRODUCTION

Wide usage of aluminum metal matrix composites in industries and while fabricating aircraft is attributed to its good strength-to-weight-to-cost ratio, as well as its excellent corrosion resistivity and non-toxicity. So, in the current study, we fabricate the composite by reinforcing it with two different reinforcements of different mixtures using friction stir processing (FSP), which is a solid-state process and also a surface composite. While there are other methods for fabricating metal matrix composites, such as liquid-state processing methods—known to be stir casting, squeeze casting, spray casting, reactive processing, and so on—casting is done by melting the metal to its liquid state and pouring it into a die to obtain the required shape and dimensions, which is then cooled to solidify the metal. However, during casting, it is observed that the cast elements are often subjected to metallurgical flaws such as porosity, microstructural issues, and casting defects. Hence, to eliminate such defects, FSP has been used to ensure a wrought microstructure.

Derived from its parent process, friction stir welding (FSW), FSP was first introduced in the year 1999. It consists of a hard pin extended by a cylindrical shoulder between the plates. FSP is used to change the properties of the material by inserting a rotating non-consumable tool into the metal piece; when it is in stirring motion with a lateral push, deformation occurs in the workpiece. Heat is also generated due to rubbing of the shoulder on the surface, which tends to soften the material and forms a solid-state bond. It is a low-heat, solid-state process. This creates a non-porous surface with a good microstructure by refining the grain structure in the solid state. FSP not only refines the grain structure, but also relieves the stresses induced while fabricating the metal plate. In FSP, we can also perform reinforcement of the materials, resulting in a surface composite.

There are several types of composites using fiber, particle, and structural reinforcement. But in FSP, we use nanoparticles or powders to be reinforced with the parent metal, so that it is a solid-state reinforcement. In this current study, we fabricate Al7075 alloy, containing the major alloying element zinc (6%), which is reinforced with boron nitride powder and multi-walled carbon nanotube (MWCNT) by using FSP. We also vary tool speed, feed, and reinforcement composition to obtain successive results. While performing FSP, we use a tool consisting of a taper pin with a circular cross-section, which is made up of high-carbon, high-chromium (HCHC) steel.

Under corrosive conditions, these joints can be attacked in various forms. Corrosion of weldments (ASM) can occur in many forms such as galvanic, pitting, stress, intergranular corrosion, hydrogen cracking, and so forth. Pitting attack occurs due to the breakdown of oxide films. For the majority of aluminum alloys, the weld metal and the HAZ become nobler relative to the base metal in a saltwater environment. Also, according to Hatch (Hatch), the aluminum alloy family (5XXX) has the lowest pitting problems in contrast to its counterpart (6XXX), where pitting and intergranular corrosion go hand in hand. So, a dissimilar joint from these two materials can have a different composition that can be highly susceptible to corrosion.

The impact of process factors on the microstructural development, mechanical characteristics, and corrosion behavior of an age-hardenable Al 7075 alloy was examined by Kumar et al. (2017). Friction stir processing (FSPed) was applied to the alloy plates with a thickness of 6 mm at different traverse speeds. It was discovered that the friction coefficient and the wear-track morphology were well-supported. Grain size refinement at greater traversal speeds is linked to an increase in mechanical characteristics (because of decreased heat input). When compared to the base metal, the FSPed samples' corrosion resistance was lower.

The research work of Kumar et al. (2020) highlights the effect of process parameters on mechanical properties of fabricated surface composites by FSP. By using various reinforcing materials like SiC, B_4C, Al_2O_3, with waste elements like waste eggshells, rice husks, coconut shell, and coir, green composites—which are environmentally friendly and reduce the problem of decomposition—were fabricated. Strength, hardness, and impact resistance are among the mechanical qualities that the reinforcement primarily improves. The outcomes showed that the mechanical characteristics of friction stir treated surface composites are highly influenced by the process parameters.

Periasamy and Sivashankar (2020) conducted an experiment on surface hybrid composites on an aluminum alloy 7075 substrate utilizing FSP and different compositions of reinforced particles, including graphite (Gr) and silicon carbide (SiC). The characteristic is better in Sample S3 (50% SiC + 50% Gr) than in other combinations. Tensile strength and hardness deteriorate with increasing Gr content in the samples.

Research by Awad et al. (2018) produced a self-lubricating AA7075/Gr composite by using the FSP technology. It was discovered that by raising the number of passes and rotation speed, the Gr particle dispersion in the AA7075 Al alloy improved. Compared to the AA7075 base alloy, the generated AA7075/Gr composite exhibited greater hardness and wear resistance.

Research by García-Vázquez et al. (2016) found that aluminum/aluminum alloy matrix composites reinforced with ceramic phases had better tribological properties and were stronger and harder than unreinforced aluminum alloys. FSP, a unique method of surface modification, has been developed to fabricate surface composites with enhanced performance. The findings showed that the surface composite's area, TiC particle distribution, and microhardness were all impacted by the chosen FSP parameters.

Madhusudhan Reddy and Srinivasa Rao (2010) established the superiority of an FSP-processed A356 when they studied and compared its corrosion behavior against an as-cast material. They found the processed A356 base material tended to have a higher pitting potential, mainly due to uniform silicon-rich eutectic in the aluminum matrix. Higher pit density was also discovered on the as-cast alloy. Rao et al. (2010) also conducted a similar study based on ASTM B-117 on gas-welded aluminum-copper alloy AA2219 and compared the results with untreated welded specimens. Unprocessed welds showed severe corrosive attack, preferentially along the inter-dendritic regions in the fusion zone and along the grain boundaries in the partially melted zone near the fusion zone. They concluded that FSP tends to eliminate copper-depleted regions in the weld microstructure and improve corrosion resistance.

Surekha et al. (2009) also studied the corrosion behavior of AA2219 friction stir processed at different RPMs and traverse speeds. They found that rotation speeds tended to influence the breakdown and dissolution of intermetallic particles, which affected the rate of corrosion. Higher speeds of rotation improved corrosion resistance, and there was no apparent impact from traverse speed.

FSWs between AA7010 and AA2024 were found by Jariyaboon et al. (2006) to increase susceptibility to corrosion due to a sensitized weld microstructure by intergranular attack in the HAZ and also in the weld zone. Changing the tool rotation speed and melting the surface with lasers was found to produce a protective layer that could offer corrosion resistance.

Kannan and Raja (2009) studied whether resistance to SCC can be improved in Al-Zn-Mg-Cu-Zr alloy by inhibiting recrystallization through the use of scandium. They conducted corrosion tests on both types of heat-treated alloys using slow strain rate testing and U-bend testing, and were able to conclude that ductile behavior continued in the alloy with scandium due to its fine grain structure, even in the peak-aged condition.

A356 alloys cast by different processes were examined by Lee et al. (2008) for their mechanical and corrosion characteristics. These, as well as microstructural features, were affected by the process used to produce them. They found the rheocasting process provided better resistance to both general and stress corrosion.

Yoshioka et al. (1986) studied the pitting corrosion behavior of rapidly solidified aluminum alloys and found that rapid solidification techniques lead to a reduction of silicon and intermetallic compound phases, except for Al-Ni-Si alloys. Verma and Misra (2021) studied several dissimilar aluminum alloy combinations welded with FSW, with respect to material positions on the advancing and retreating sides of the tool, mechanical properties, and chemical composition using EDS. They found that the key element from the advancing-side material tended to be of a higher weight percentage in the weld zone. Moreover, tensile test fractures were located on the side of the material with lower strength.

The above-mentioned research suggests FSP's ability to improve the corrosion resistance of aluminum alloys. Therefore, the overall objective is to fabricate Al7075 reinforced with boron nitride and carbon nanotube (CNT) using FSP and to study its mechanical properties. Using specimens with varied mixtures of reinforcements, a study is carried out on the particle reinforcement properties of BN and CNT with the Al7075 metal matrix and the mechanical properties of the specimens.

7.2 EXPERIMENTS AND RESULTS

The alloys used for the experimentation are AA7075, BN, and MWCNT. While performing FSP on a metal, tool rotation speed and feed play an important role by generating heat in the workpiece due to the friction created by the shoulder. The tool speed and feed are responsible for mixing nanoparticles (reinforcements) in the workpiece to obtain a uniform composite. So, it must be ensured to fix the optimum tool rotation speed and feed to achieve good results. The experiments are carried out with three different tool rotation speeds while keeping the feed rate constant, such that the optimal speed is obtained. Further, various combinations of the feed rate with rotation speed were followed as mentioned in Table 7.1.

TABLE 7.1 Selected Tool Rotation Speeds and Feed Rates

SL. No.	Tool Rotation Speed (rpm)	Tool Travelling Feed (mm/min)
1.	1000	20
2.	1500	20
3.	2000	20
4.	1500	30
5.	1500	40

The 7075 alloy was a good choice for the welding process due to its better corrosion properties. After performing the experiment with various tool rotation speeds and feed rates, hardness and corrosion tests were conducted so that the optimum tool rotation speed and feed rate could be determined for performing FSP reinforced with hexagonal boron nitride and CNTs.

Workpieces made of AA7075, in the shape of a block sized 150 × 100 × 8 mm, were prepared for the processing. The FSP process was performed on a CNC Mill MCV-450 (Make: ACE Micromatic) available in-house, as shown in Figure 7.1(a). The welding tool, made of HCHC steel, was designed and manufactured and is also shown in Figure 7.1(c). The HCHC tool is a tapered cylindrical shank with a shoulder diameter of 20 mm and a conical pin (pin length 3.2 mm, pin diameter 6–5 mm, shoulder diameter 18 mm). It is run at 1000–2000 rpm with a translational speed of 20, 30, and 40 mm/min. Welding was performed on the CNC Milling Machine MCV-450, equipped with a Fanuc Controller capable of spindle speeds up to 10,000 rpm and a table measuring 800 × 450 × 500 mm.

The process of fabricating the composite involved placing and fixing the block on the table, positioning the tool, making the presets on the machine tool control panel, and initiating the weld run. The tool is passed through only once during each welding operation. While performing nanoparticle reinforcement with a metal using FSP, small grooves of 3 mm depth and 2.5 mm diameter were made to fill the nanoparticles, so that the packing of reinforced particles in the groove could be achieved with the help of a pinless tool. After packing the reinforcement particles with the pinless tool, FSP was performed using a tapered cylindrical profile tool. The experiments were carried out with three different tool rotation speeds while keeping the feed rate constant, so that the optimal speed could be determined (Figure 7.2). Further, the feed rates were varied while keeping the optimal rotation speed fixed, as shown in Table 7.1.

After performing the experiment with various tool rotation speeds and feed rates, hardness and corrosion tests were conducted so that the optimum tool rotation speed and feed rate could be set for performing FSP reinforced with hexagonal boron nitride and CNTs.

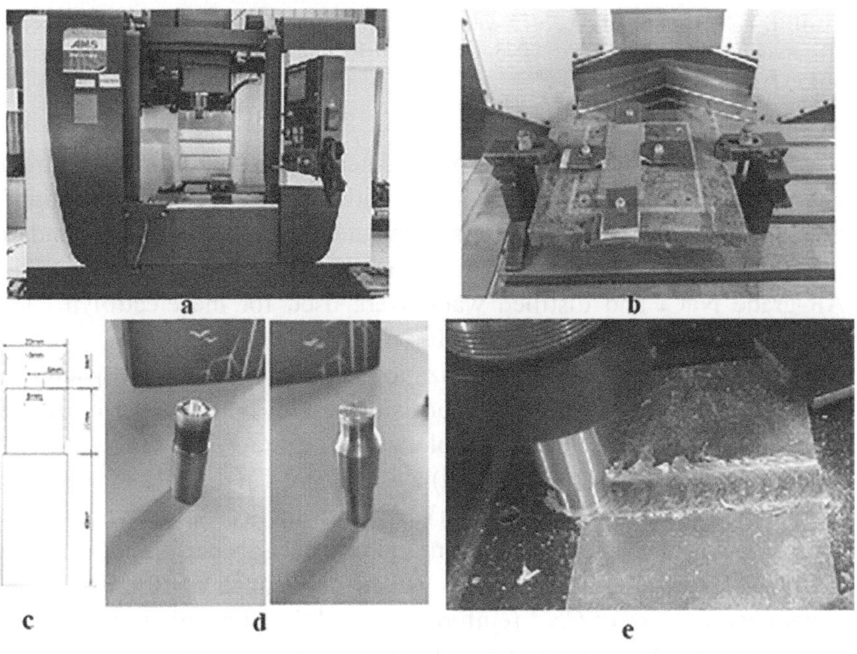

FIGURE 7.1 (a) FSW setup; (b) worktable; (c) cylindrical pin profile; (d) fabricated FSP tools (tapered cylindrical pin and pinless); and (e) FSP pass on AA7075 reinforced with h-BN and MWCNTs.

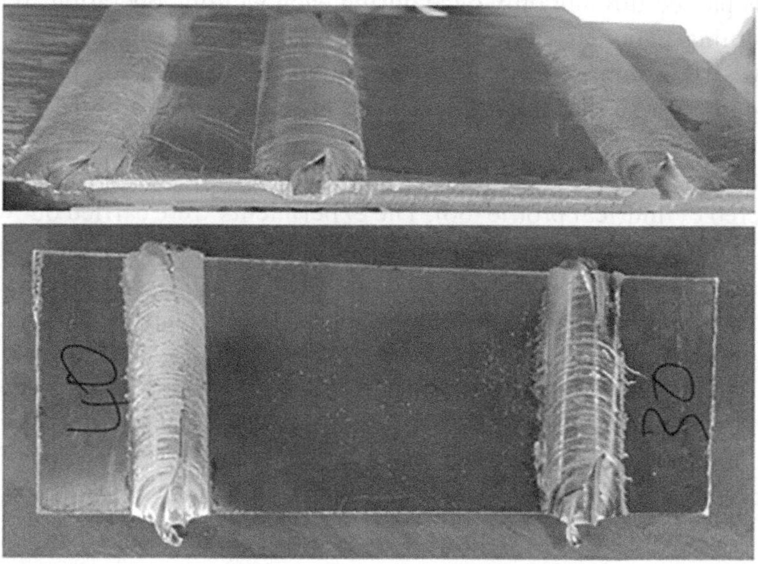

FIGURE 7.2 (a) FSP of AL 7075 with 1000, 1500, and 2000 rpm at a constant feed of 20 mm/min; and (b) FSP of AL 7075 with 30 and 40 mm/min at a constant rotation speed of 1500 rpm.

Pitting corrosion susceptibility was measured using a three-electrode configuration on electrochemical workstation CHI-608E by conducting a potentiodynamic polarization test. The reference electrode is a standard calomel electrode (SCE), and the counter electrode is a carbon electrode. AR-grade NaCl and distilled water were used for the electrolyte. The potential scan was carried out at 0.01 V/s for all specimens.

As seen in Figure 7.3, the polarization outputs for all FSPed specimens were obtained from the pitting corrosion test by varying the corrosion potential. From the Tafel curves plotted in Figure 7.3, a higher passive potential (or less negative potential) was observed in the case of specimen A (blue-colored curve) at –681.7 mV, which is the least negative potential compared to other specimens. From the comparison bar graph in Figure 7.4, it was seen that at a constant weight percentage of BN, the lowest weight percentages of CNT reinforcement exhibited the highest corrosion resistance of AA7075. The fine nanosized particles of CNT and micro-sized particles of BN in the aluminum metal matrix enhance the number of insulation paths between the anodic matrix and cathodic intermetallic locations, leading to better corrosion resistivity.

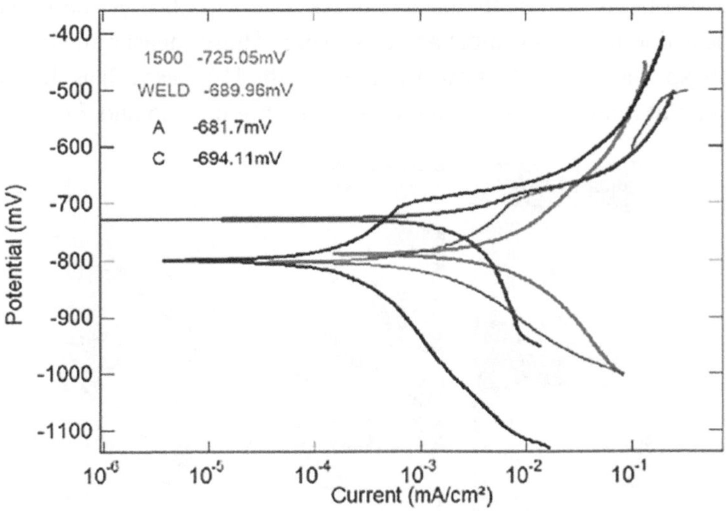

FIGURE 7.3 Tafel curves for AA7075 reinforced with BN and CNT mixtures using FSP.

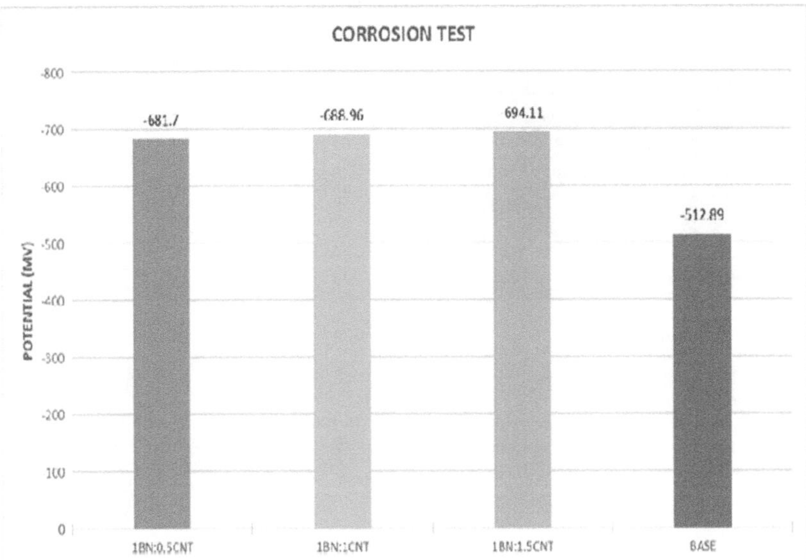

FIGURE 7.4 Comparison bar graph of corrosion test on AA7075 reinforced with BN and CNT mixtures.

Hardness tests on the polished surface across the FSP zone were performed using a Rockwell hardness tester at 100 Kgf. Hardness was measured across the FSP zone as well as the base material zones to observe the variation in hardness due to thermo-mechanical stirring. These measurements, taken at equally spaced intervals, show the extent of the HAZ as well as the variation from parent material to the FSPed zone (*see* Figures 7.5 and 7.6).

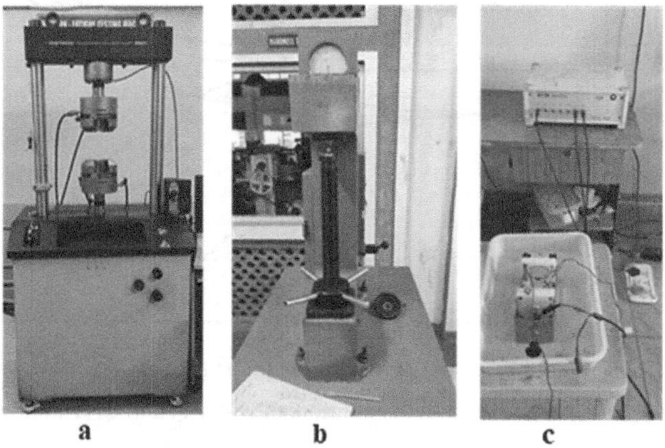

FIGURE 7.5 (a) UTM; (b) Rockwell hardness tester; and (c) pitting corrosion tester.

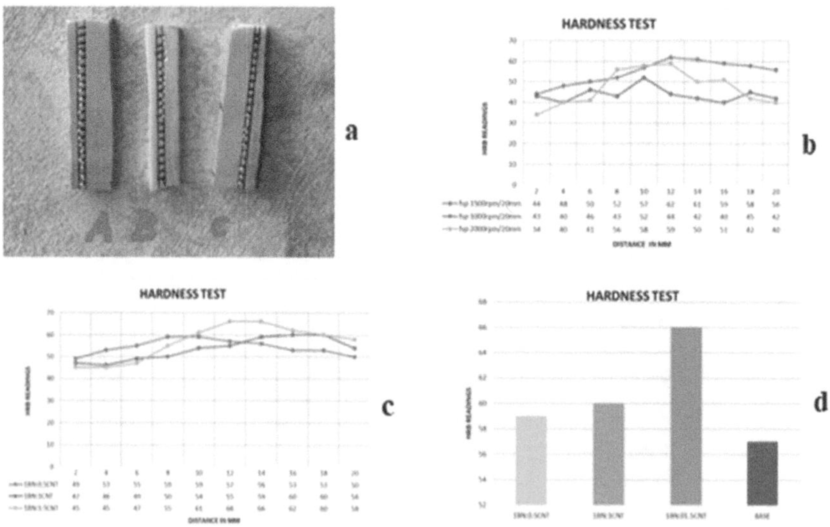

FIGURE 7.6 (a) Specimen used for hardness test, with indenter point distances noted for FSP of varying mass ratios, where A, B, and C represent 1BN:0.5CNT, 1BN:1CNT, and 1BN:1.5CNT; (b) Rockwell test graph of FSP metal with varying tool rotation speed and feed; (c) Rockwell readings obtained for FSP metal with varying BN and CNT weight ratios; and (d) Rockwell bar graph obtained for FSP metal with varying BN and CNT weight ratios.

Subsequently, the specimens shown in Figure 7.7(a) were subjected to a standard tensile test in a universal testing machine to measure the strength across the FSPed zone. In spite of the dissimilar zones in the specimen, the test is primarily used to determine whether there is any variation in strength due to FSP (*see* Figure 7.7).

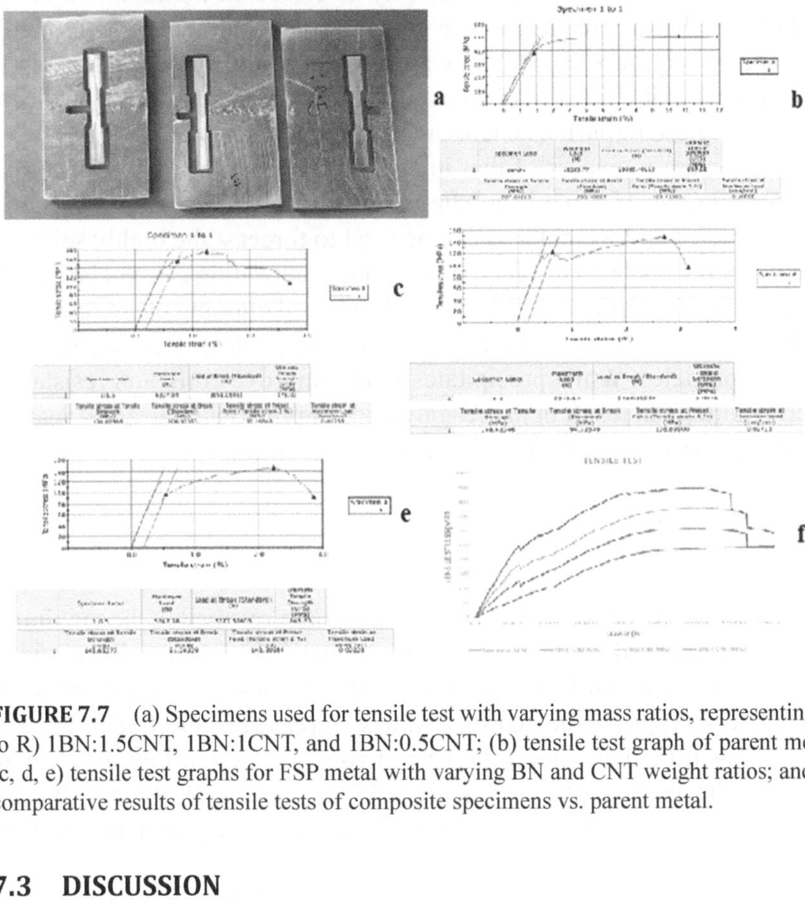

FIGURE 7.7 (a) Specimens used for tensile test with varying mass ratios, representing (L to R) 1BN:1.5CNT, 1BN:1CNT, and 1BN:0.5CNT; (b) tensile test graph of parent metal; (c, d, e) tensile test graphs for FSP metal with varying BN and CNT weight ratios; and (f) comparative results of tensile tests of composite specimens vs. parent metal.

7.3 DISCUSSION

7.3.1 CORROSION ANALYSIS

Tafel curves showing pitting susceptibility of the base metal as well as FSPed workpieces are shown in Figure 7.3. The corrosion resistance values for AA7075 were observed to be lower than those of composited AA7075. It has been proved in earlier studies that corrosion can be influenced by the presence of micro-constituents—their size, quantity, location, and corrosion potential relative to the adjacent aluminum solid solution.

The corrosion resistance of alloys can be influenced by alloying elements, contingent on their presence as second-phase precipitates dispersed throughout the matrix, separated at grain boundaries, or present in solid solution. Iron- or silicon-based micro-constituents can be cathodic to the aluminum matrix. The possibility of corrosion in aluminum alloys containing silicon percentages near eutectic composition in aqueous media increases when these constituents are present at the surface with thin or non-existent oxide films over them. In a conductive liquid, the local cells created by these impurities encourage surface pitting attacks.

Magnesium in excess of that in solid solution in the alloy can form a component (Mg_2Al_3) that is anodic to the Al-Mg solid solution. Potential measurements and weight loss can be used to forecast the highly selective attack that results from the anodic component corroding electrochemically in a corrosive environment. Limited solubility of alloying elements results in their presence in solid solution, in fine precipitates, or as coarse intermetallic particles. While precipitates might improve corrosion resistance, intermetallics can result in micro-galvanic coupling with the alloy, leading to pitting corrosion as well as intergranular cracking.

7.3.2 HARDNESS

It was observed that the 1BN:1.5CNT mass ratio resulted in the maximum hardness compared to the base metal. As the CNT content increased in the reinforcement, hardness also increased. Hardness measurements made across the workpiece showed that values increased in the FSPed zone as well as toward either side of the FSPed zone, as seen in Figure 7.6. This indicates the absence of a significant HAZ on either side of the FSP zone. It also suggests that heat generation did not cause any abnormally high temperature in the stir area.

At any point in time, the stirred weld material directly under the tool shoulder undergoes varying degrees of movement. Material nearer the center of the tool undergoes the least movement and is therefore relatively less dynamic. The temperature in this zone is also the highest compared to the stir zone's outer edges. The plot of hardness values against distance measured from the leftmost point shows a peak in hardness at the center of the weld zone, approximately where the pin on the tool is located. The stir zone is assumed to be 20 mm wide, which might be the same as the heat-affected zone.

7.3.3 TENSILE TESTS

The outputs from the tensile tests are shown in Figure 7.7. The tensile strength of the processed specimen was observed to be approximately 176.03 MPa. Compared to the parent metal's zone, this value was significantly lower than 507.05 MPa. Possible reasons could be that the processed zone underwent changes in grain structure due to dynamic recrystallization, resulting from vigorous stirring and plastic flow of the heated materials. The lower heat input, compared to other welding methods, along with air cooling, likely resulted in gradual heating and cooling, which may have led to the formation of a fine-grained structure. This change in grain size in the weld zone might have contributed to the reduced strength.

Since no PWHT was carried out after the welding process, improved tensile properties cannot be expected to exceed those of the parent materials, which were procured in a specific tempered condition. Ideally, the surface composite should show strength above that of the parent metal. However, as the processed workpiece is weaker, some PWHT may be required.

7.4 CONCLUSION

Considering the results from the corrosion study, the weld joint exhibits lower susceptibility to pitting corrosion compared to the parent materials. The alloying elements present in the weld zone are likely responsible for the improved corrosion resistance. Overall, the weldability of the two alloys considered appears to be sufficiently high, as indicated by the relatively high weld joint strength, although it remains lower than that of the parent materials.

The hardness values also help in determining the length of the HAZ, which suggests that there was no undue influence of heat from the welding process on the materials being joined. The increased hardness is likely the result of the presence of alloying elements in the stir zone, whereas tensile strength can only be improved by age hardening. The homogeneity of the weld and the absence of any visible defects indicate that the two materials are fairly compatible.

ACKNOWLEDGMENTS

The authors are thankful to Raghu Engineering College for providing their facilities and their continued encouragement to publish this work.

DATA AVAILABILITY

The experimental data generated during this research are presented throughout this article and constitute the raw data available within this work. The current research also includes both raw and processed data required to replicate these results.

KEYWORDS

- **corrosion**
- **friction stir processing**
- **micro-constituents**
- **post weld heat treatment**
- **surface nano-composite**
- **tensile test**
- **thermo-mechanical stirring**
- **universal testing machine**

REFERENCES

Awad, O. M., Seleman, M. M., Ahmed, M. M. Z., & Ammar, H. R. (2018). Production and characterization of AA7075-graphite composite using friction stir processing. *Journal of Petroleum and Mining Engineering*, *20*(1), 101–110.

García-Vázquez, F., Vargas-Arista, B., Muñiz, R., Ortiz, J. C., García, H. H., & Acevedo, J. (2016). The role of friction stir processing (FSP) parameters on TiC reinforced surface Al7075-T651 aluminum alloy. *Soldagem & Inspeção*, *21*, 508–516.

Jariyaboon, M., Davenport, A. J., Ambat, R., Connolly, B. J., Williams, S. W., & Price, D. A. (2006). Corrosion of a dissimilar friction stir weld joining aluminum alloys AA2024 and AA7010. *Corrosion Engineering, Science and Technology*, *41*(2), 135–142.

Kannan, M. B., & Raja, V. S. (2010). Enhancing stress corrosion cracking resistance in Al–Zn–Mg–Cu–Zr alloy through inhibiting recrystallization. *Engineering Fracture Mechanics*, *77*(2), 249–256.

Kumar, A., Sharma, S. K., Pal, K., & Mula, S. (2017). Effect of process parameters on microstructural evolution, mechanical properties and corrosion behavior of friction stir processed Al 7075 alloy. *Journal of Materials Engineering and Performance*, *26*, 1122–1134.

Kumar, S., Srivastava, A. K., & Singh, R. K. (2020, August). Fabrication of AA7075 hybrid green metal matrix composites by friction stir processing. In *Annales de Chimie-Science des Matériaux* (Vol. 44, No. 4, pp. 295–300).

Lee, K., Kwon, Y. N., & Lee, S. (2008). Correlation of microstructure with mechanical properties and fracture toughness of A356 aluminum alloys fabricated by low-pressure-casting, rheo-casting, and casting–forging processes. *Engineering Fracture Mechanics*, *75*(14), 4200–4216.

Madhusudhan Reddy, G., & Srinivasa Rao, K. (2010). Enhancement of wear and corrosion resistance of cast A356 aluminum alloy using friction stir processing. *Transactions of the Indian Institute of Metals*, *63*, 793–798.

Periasamy, K., Sivashankar, N., Chandrakumar, S., & Viswanathan, R. (2020). Measurement of friction and wear in aluminum alloy Al7075/Sic & Gr processed by friction stir method. *Int. J. Innov. Technol. Explor. Eng*, *9*, 278–281.

Rao, K. P., Ram, G. J., & Stucker, B. E. (2010). Effect of friction stir processing on corrosion resistance of aluminum–copper alloy gas tungsten arc welds. *Materials & Design*, *31*(3), 1576–1580.

Surekha, K., Murty, B. S., & Rao, K. P. (2009). Effect of processing parameters on the corrosion behavior of friction stir processed AA 2219 aluminum alloy. *Solid State Sciences*, *11*(4), 907–917.

Verma, S., & Misra, J. P. (2021). Experimental investigation on friction stir welding of dissimilar aluminum alloys. *Proceedings of the Institution of Mechanical Engineers, Part E: Journal of Process Mechanical Engineering*, *235*(5), 1545–1554.

Yoshioka, H., Yoshida, S., Kawashima, A., Asami, K., & Hashimoto, K. (1986). The pitting corrosion behavior of rapidly solidified aluminum alloys. *Corrosion science*, *26*(10), 795–812.

CHAPTER 8

Role of Slag Viscosity in the Blast Furnace Process of Pig Iron Production and Its Estimation

S. BEURA,[1,*] U. K. MOHANTY,[1,2] P. K. SATAPATHY,[1] and S. N. PATI[1]

[1]*Department of Mechanical Engineering, ITER, SOA (Deemed to be) University, Bhubaneswar, Odisha, India*

[2]*Department of Metallurgical and Materials Engineering, NIT Rourkela, Odisha, India*

*Corresponding author

ABSTRACT

Together, CaO, SiO_2, MgO, and Al_2O_3—the four main components of BF slag—are responsible for determining the slag viscosity. The slag viscosity can be either low or high in high-Al_2O_3 slags, depending on the Al_2O_3/CaO and/or Al_2O_3/SiO_2 ratio, which has a substantial impact on the BF route's iron production process. To provide sufficient data in this regard, the current work intends to analyze the viscosity data for high-Al_2O_3 BF slag as impacted by the presence of the main elements CaO, SiO_2, and MgO.

8.1 INTRODUCTION

Slag viscosity is one of the most crucial variables in the investigation of blast furnace (BF) operation, as it is governed by its composition and affected by its structure. The trapping of the expensive hot metal in the seemingly pointless slag is controlled by the viscosity of the BF slag, which regulates slag-metal separation in the BF. In addition, the furnace's heat

Advances in Industrial Manufacturing and Materials: Integrating a New Era of Computational Modeling and Machine Learning. Debesh Mishra and Swagatika Mishra (Eds.)

DOI: 10.1201/9781779640475-8

transfer, the life of the refractory lining, and—more importantly—the slag-metal reactions and exchanges are all direct functions of BF slag viscosity.

Al_2O_3, owing to its complicated chemical behavior in the presence of other major constituents—CaO, SiO_2, and MgO—in the slag, is the greatest concern in the study of BF slag viscosity. Al_2O_3 adopts a four-fold (AlO_4) or a six-fold (AlO_6) coordination in silicate melts, acting as both a network former and a network modulator (Weyl, 1951). Whether four or six oxygen ions are necessary to screen the Al^{3+} ion depends on the presence of other cations in the slag melt. Al_2O_3 adopts a four-fold coordination, and the melt contains SiO_4^{4-} and AlO_4^{5-} ions when the molar ratio of Al_2O_3/CaO is less than unity—that is, when sufficient basic oxide is present in the slag melt (Machin et al., 1952). In this situation, Al_2O_3 in the silicate melt serves as a network former.

Al, on the other hand, adopts a six-fold coordination and penetrates the interstices in the structure, functioning as a network modifier when sufficient oxygen is not present for the formation of Al and Si tetrahedra—that is, when the Al_2O_3/CaO ratio is greater than unity (Moore and McMillan, 1956). In other words, Al_2O_3 and SiO_2 react similarly in neutral and basic slags when the Al_2O_3/CaO ratio is less than or equal to unity. Here, the lattice positions of Al and Si are comparable, and the ions that make up the network are (Al + Si).

Al_2O_3 acts as a network breaker to reduce the viscosity of the slag when both the Al_2O_3/CaO ratio and SiO_2 content are high. As a result, Al_2O_3 is classified as an amphoteric oxide.

Due to their strong covalent bonds, SiO_2 and Al_2O_3 make the BF slag more viscous. The silicate network is destroyed, and the viscosity is reduced by monoxides like CaO and MgO, on the other hand. However, when the monoxide activity increases, solid phases like *gehlenite* (calcium aluminum silicate; melting point: 1590°C) precipitate, which raises the slag melt's overall viscosity (Sunahara et al., 2008).

In a silicate slag melt, in addition to the discrete silicate ions, singly bonded oxygen (O^-), doubly bonded oxygen (O^0), and free oxygen (O^{2-}) are always present. Thus, in silicate melts containing Al-Si, depending on the Al_2O_3/SiO_2 ratio—or simply the Al/Si ratio—Al^{3+}–O^- type reactions are a strong probability. The rate of Al^{3+}–O^- interactions is enhanced at higher Al_2O_3/SiO_2 ratios, that is, at higher Al/Si ratios. These interactions increase the degree of polymerization in the silicate melt as influenced by the presence of other oxides. As a consequence, a general tightening of the ionic bonds occurs, and the activation energy for viscous flow of the melt increases, thereby increasing the slag viscosity.

A combination of high MgO and high C/S ratio in the silicate melt results in a short slag (Dash et al., 2012; Gupta and Sheshadri, 1973; Singh, 1984). The formation of such a slag is beneficial from the standpoint of slag flow characteristics, as it can attain free-flowing behavior after melting without the need for substantial heat input. However, from the viscosity point of view, such a slag may cause problems if the operating temperature falls below a certain threshold value.

Thus, it can be appreciated that the major constituents of BF slag—viz., CaO, SiO_2, MgO, and Al_2O_3—are jointly responsible for influencing slag viscosity. In high-Al_2O_3 slags in particular, depending on the Al_2O_3/CaO and/or Al_2O_3/SiO_2 ratio, the slag viscosity can either be low or high, having a significant influence on the BF route's iron production process. The present work, therefore, aims to analyze the viscosity data for high-Al_2O_3 BF slag as influenced by the presence of the major constituents CaO, SiO_2, and MgO to generate sufficient data in this respect.

8.2 EXPERIMENTAL WORK

8.2.1 PREPARATION OF SYNTHETIC SLAG IN THE LABORATORY

Synthetic slag samples (Tables 8.1–8.3) were prepared in the laboratory using analytical-grade oxides with a purity ranging from 98.50 to 99.90%. The compositions of these slags were selected with consideration of the slag compositions typically encountered in different operating BFs in the country.

TABLE 8.1 Chemical Compositions (in mass %) of the Slag Used for the C/S Variation Investigation

Sl. No.	Al_2O_3	MnO	K_2O	Na_2O	Fe_2O_3	TiO_2	MgO	CaO	SiO_2	CaO/SiO_2 (C/S)
1.	20.00	00.10	00.50	01.10	01.00	00.60	08.00	32.54	36.16	00.90
2.	20.00	00.10	00.50	01.10	01.00	00.60	08.00	33.47	35.23	00.95
3.	20.00	00.10	00.50	01.10	01.00	00.60	08.00	34.35	34.35	01.00
4.	20.00	00.10	00.50	01.10	01.00	00.60	08.00	35.19	33.51	01.05
5.	20.00	00.10	00.50	01.10	01.00	00.60	08.00	35.99	32.71	01.10

TABLE 8.2 Chemical Compositions (in mass %) of the Slag That Was Used for the MgO Variation Investigation

Sl. No.	Al_2O_3	MnO	K_2O	Na_2O	Fe_2O_3	TiO_2	CaO	SiO_2	CaO/ SiO_2 (C/S)	MgO
6.	20.00	00.10	00.50	01.10	01.00	00.60	36.35	36.35	01.00	04.00
7.	20.00	00.10	00.50	01.10	01.00	00.60	35.35	35.35	01.00	06.00
8.	20.00	00.10	00.50	01.10	01.00	00.60	34.35	34.35	01.00	08.00
9.	20.00	00.10	00.50	01.10	01.00	00.60	33.35	33.35	01.00	10.00
10.	20.00	00.10	00.50	01.10	01.00	00.60	32.35	32.35	01.00	12.00

TABLE 8.3 Chemical Compositions (in mass %) of the Slag Chosen for the Investigation of the Simultaneous Fluctuation in the C/S Ratio and MgO for Measuring Viscosity

Sl. No.	Al_2O_3	MnO	K_2O	Na_2O	Fe_2O_3	TiO_2	CaO	SiO_2	CaO/ SiO_2 (C/S)	MgO
11.	20.00	00.10	00.50	01.10	01.00	00.60	34.44	38.26	00.90	04.00
12.	20.00	00.10	00.50	01.10	01.00	00.60	34.44	36.26	00.95	06.00
13.	20.00	00.10	00.50	01.10	01.00	00.60	34.35	34.35	01.00	08.00
14.	20.00	00.10	00.50	01.10	01.00	00.60	34.16	32.54	01.05	10.00
15.	20.00	00.10	00.50	01.10	01.00	00.60	33.89	30.81	01.10	12.00

The oxides are weighed in the given proportions (according to the chemical details of the industrial slag), taken in a bottle, and shaken well for about one hour with the lid secured in position. For homogenization, the sample is mixed in a planetary ball mill in a tungsten carbide (WC) bowl using WC balls for 24 hours. The ground sample is heated in a raising hearth furnace at 1823 K in air in a Pt crucible and held for 2 hours at that temperature to form the slag melt. This melt is quenched in water and solidifies to a glassy mass. The glassy slag is crushed and ground in the planetary ball mill for 6 hours. The ground sample is then re-melted in the raising hearth furnace as described earlier. This time, the holding time at temperature is 4 hours, with intermittent stirring using a Pt-Rh rod for homogenization. These molten samples are quenched, dried, and again ground in the planetary ball mill for 2 hours.

Chemical analysis of the ground samples shows insignificant compositional variation compared to the initially weighed proportions of the oxides. These samples, with compositional details presented in Tables 8.1–8.3, are used for experimentation.

8.2.2 CHEMICAL COMPOSITION OF THE SYNTHETIC SLAG

As shown in Tables 8.1–8.3, the synthetic slags are classified into three groups, each consisting of five slag samples. The first group focuses on C/S ratio fluctuation, the second group on MgO content variation, and the third group on simultaneous variation of the C/S ratio and MgO content.

In the first batch of slag samples, the C/S ratio is changed in 0.05-step increments from 0.9 to 1.1. MgO is varied in the second group in increments of 2 mass %, ranging from 4 mass % to 12 masses %. In the third group of slag samples, MgO is varied from 4 to 12 mass % in stages of 2 mass %, while the C/S ratio is changed from 0.9 to 1.1 in steps of 0.05. The variations are arranged so that both variables increase and decrease concurrently: MgO reaches its minimum value of 4 mass % when the C/S ratio is at its lowest (0.9), and MgO reaches its maximum value of 12 mass % when the C/S ratio is at its highest (1.1).

Al_2O_3, MnO, K_2O, Na_2O, Fe_2O_3, and TiO_2 are maintained at 20, 0.10, 1.10, 0.50, 1.0, and 0.6 mass %, respectively, in each slag sample. The percentage composition of the other variables is held at the midpoint in slag samples intended for evaluating the impact of a single component. For example, MgO is fixed at 8 mass % in samples designed for C/S ratio investigation, and the C/S ratio is fixed at 1.0 in slags intended for MgO variation research.

8.2.3 MEASUREMENT OF SLAG VISCOSITY

8.2.3.1 HIGH TEMPERATURE VISCOMETER

Figure 8.1 provides a schematic representation of the high-temperature viscometer, model VIS 403 HF, manufactured by BAHR, Germany, which was utilized in this study to evaluate the viscosity of slag. The dynamic viscosity of this type of inner-cylinder rotating viscometer is determined by measuring a rotor's torque while submerged in the slag melt.

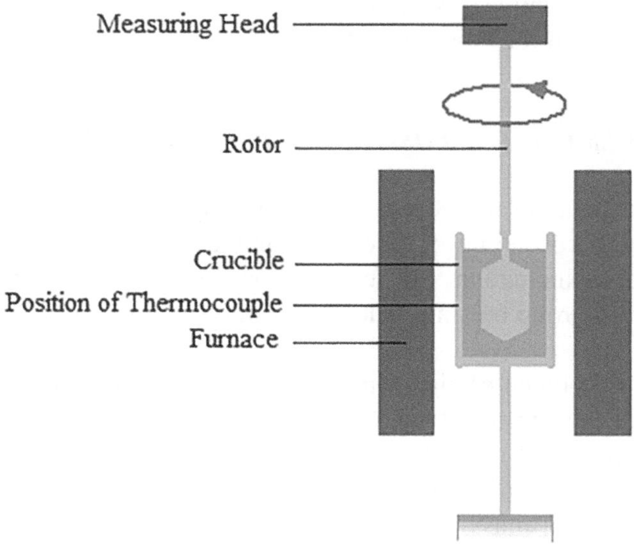

FIGURE 8.1 Schematic diagram of the model VIS 403 HF high-temperature viscometer.

8.2.3.2 VISCOMETER'S CALIBRATION

Prior to experimentation, the viscometer is calibrated using standard glass granules of known viscosity at specific temperatures, as supplied with the instrument. The measured data for the standard sample, within the range of temperatures at which slag viscosity is to be measured, are compared with the standard values provided by the manufacturer. From the difference between the standard and measured values, a correction factor is determined. This correction factor is then incorporated into the machine. Once entered, the system automatically records the adjusted measured values.

This correction factor is periodically reviewed and updated by incorporating a new correction factor into the machine at regular intervals.

Initially, a standard slag (40 mass % CaO, 40 mass % SiO_2, and 20 mass % Al_2O_3) is synthetically prepared in the laboratory, for which viscosity versus temperature data are available in the literature (Saito et al., 2003). The viscosity of this slag is measured using the standard procedure. Once the measured data are found to be comparable with the published literature data, the machine is considered ready for use. Table 8.4 shows that the deviation between the measured data and the literature values is a maximum of 5%.

TABLE 8.4 Comparison of Viscosity Data for the Slag Containing 40 mass % CaO, 40 mass % SiO_2, and 20 mass % Al_2O_3

Slag No.	Temperature (K)	Viscosity (P) (Measured)	Viscosity (P) (Literature)	Percentage Difference (%)
1.	1678	21.96	22.53	2.60
2.	1718	14.50	15.19	4.20
3.	1758	09.79	10.28	5.00

8.3 RESULTS AND DISCUSSION

8.3.1 VARIATION OF C/S RATIO

As shown in Figure 8.2, the observed viscosity values for Slag Nos. 1 to 5—intended for the investigation of the impact of changes in the C/S ratio—are plotted versus temperature. The slopes of the figure make it clear that temperature is the most significant factor in reducing viscosity for the slag with the lowest C/S ratio. As the C/S ratio increases, the slope of the curve decreases, and the rate at which viscosity decreases with rising temperature becomes slower.

The lowering of viscosity may result either from the weakening of ionic bonds between the cations (Ca^{2+}) and silicate anions due to increased temperature, or from the breakdown of the silicate network into smaller flow units caused by depolymerization due to the addition of cations (Ca^{2+}) (Mohanty, 1998). When the C/S ratio is low—that is, when CaO content is low—its ability to break down the silicate network into smaller flow units is also limited. In such cases, the temperature rise contributes more significantly to the reduction in viscosity than the limited network-modifying effect of CaO.

This is also evident from Figure 8.4, which shows straight-line plots between viscosity and temperature, indicating that the viscosity decrease is primarily due to the weakening of the bonds between Ca^{2+} cations and discrete silicate anions, rather than from increased formation of smaller flow units. In this case, the degree of depolymerization remains practically constant. Hence, at lower C/S ratios, temperature rise plays a greater role in reducing viscosity.

On the other hand, when the C/S ratio is high—that is, when CaO content is high—the decrease in viscosity results from the breakdown of the silicate network into smaller flow units, indicating a greater degree of depolymerization.

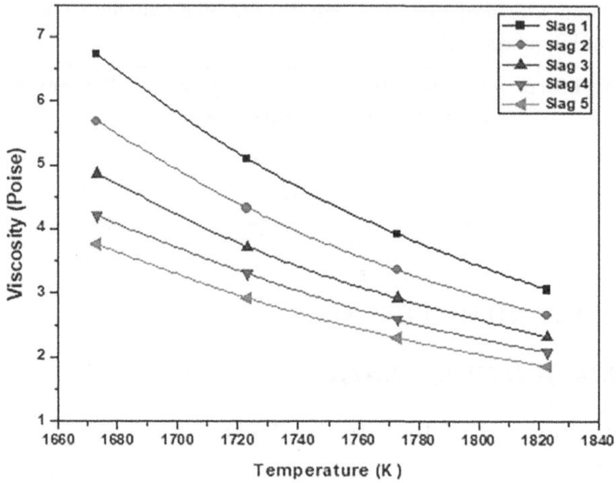

FIGURE 8.2 Viscosity variations with temperature at different C/S ratios.

The viscosity isotherms at various temperatures as a function of mol % CaO are shown in Figure 8.3. It is intriguing to observe how the graphs gradually flatten with increasing temperature, eventually becoming nearly straight lines around 1823 K. This indicates that CaO has a limited capacity to reduce viscosity at higher temperatures. Based on this observation, it can be concluded that at lower temperatures, CaO has a greater influence on altering the anionic silicate structure. Therefore, it is advised to carefully control the C/S ratio in the slag when operations are conducted at comparatively lower temperatures.

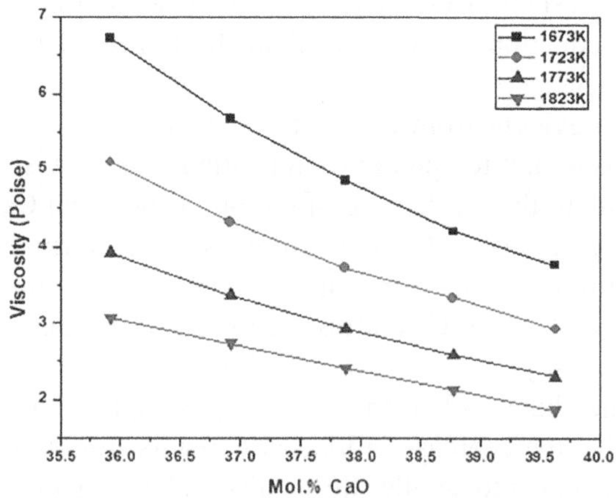

FIGURE 8.3 Viscosity variations with mol % CaO.

The straight-line relationship between the activation energy of viscous flow (E_μ) and the inverse of temperature ($1/T$) establishes that slag viscosity decreases with increasing temperature, and that the variation of viscosity with temperature follows the Arrhenius equation, $\mu = A_0 \exp(-E_{(\mu)}/RT)$ (Figure 8.4). The slope of any plot in Figure 8.4, which represents the E_μ value for a particular slag, remains constant. This demonstrates that for any given slag composition at different temperatures, the degree of polymerization is nearly constant. Therefore, it can be concluded that the discrete anionic units do not fragment into smaller flow units with increasing temperature and that the observed decrease in viscosity is solely due to the weakening of ionic bonds between Ca^{2+} cations and discrete silicate anions.

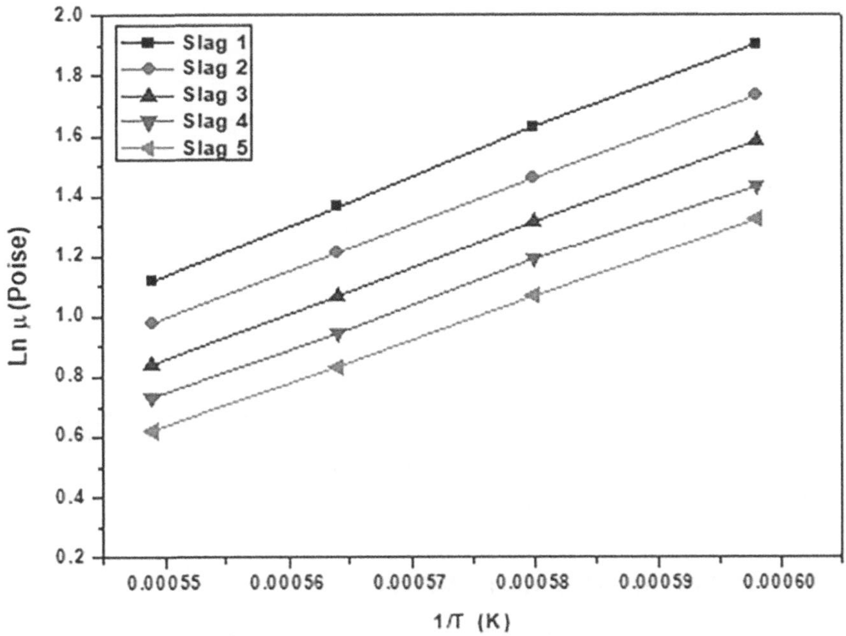

FIGURE 8.4 Variation of $\ln(\mu)$ with $1/T$.

In Figure 8.5, the E_μ values computed from the slopes of the plots in Figure 8.4 are shown as a function of the mole percent CaO. The charts clearly demonstrate that as the CaO mol % increases, the $E\mu$ values steadily decline. This confirms that increasing CaO content leads to the breakdown of the silicate network structure into more depolymerized flow units, resulting in the formation of increasingly smaller structural units.

Additionally, the rate of decline of E_μ becomes greater at higher CaO mole percentages, as reflected by the steeper slope changes. This occurs because the metallic cations Ca^{2+} supply more oxygen when the anionic silicate flow units are smaller, thus increasing the oxygen demand. As a result, Tiwari et al. (2013) observed that gradual addition of CaO becomes progressively less efficient at inducing silicate network depolymerization.

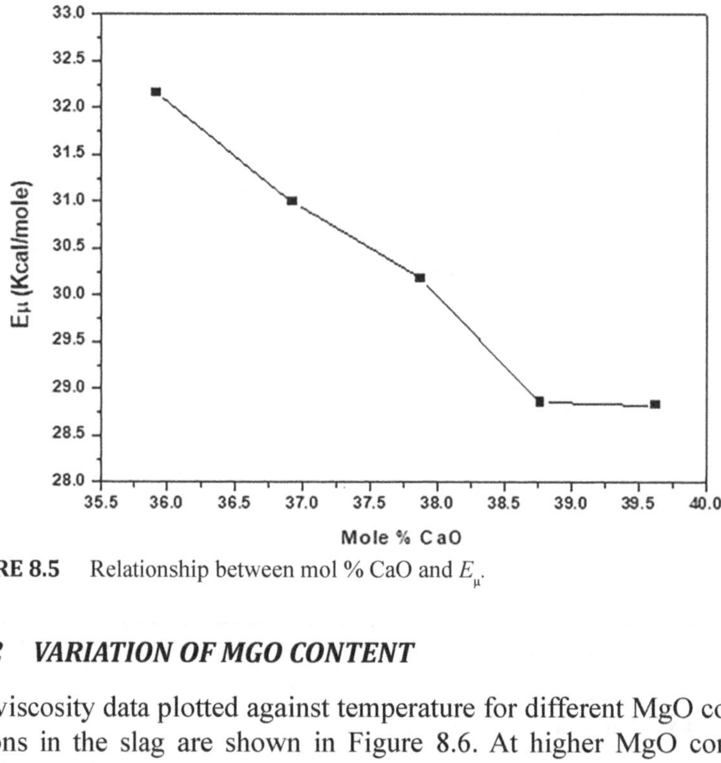

FIGURE 8.5 Relationship between mol % CaO and E_μ.

8.3.2 VARIATION OF MGO CONTENT

The viscosity data plotted against temperature for different MgO concentrations in the slag are shown in Figure 8.6. At higher MgO contents, temperature increases have a diminishing effect on reducing viscosity. Viscosity also decreases with increasing MgO concentration. This trend is similar to the behavior observed with variation in the C/S ratio.

As previously noted, MgO functions as a network modifier. At lower MgO levels, its contribution to breaking down the silicate network is limited. In these cases, the observed reduction in viscosity is attributed more significantly to temperature, which weakens the bonding between discrete anions and cations.

Furthermore, it has been found that, regardless of temperature variation, the rate of viscosity decrease is greater at lower MgO levels than at higher MgO contents. This is because, beyond approximately 10 mass % MgO, the excess MgO does not incorporate into the slag structure in a stoichiometric manner. Instead, it leads to the formation of a spinel phase composed of magnesium and aluminum (MgO–Al_2O_3), which crystallizes at relatively low process temperatures (melting point: 2288 K). This crystallization slows the reduction in viscosity.

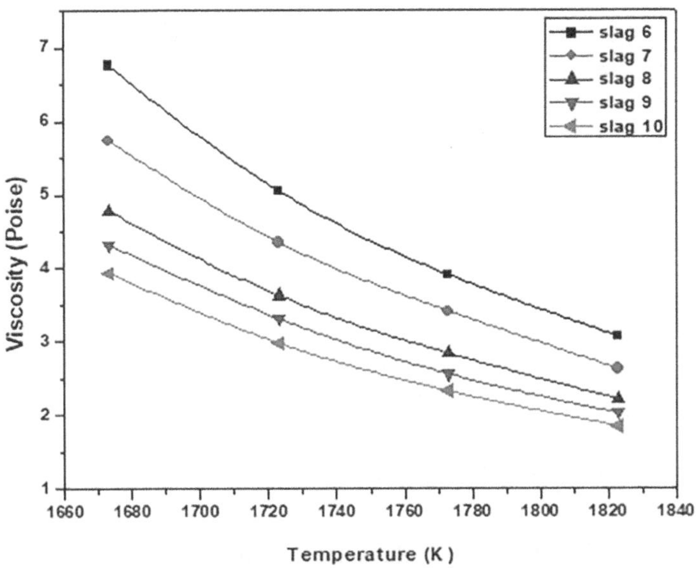

FIGURE 8.6 Viscosity vs. temperature as a function of MgO variation.

The viscosity isotherms as a function of MgO mol % at different temperatures are presented in Figure 8.7. The viscosity of the slag decreases with increasing temperature across all levels of MgO content and temperature ranges investigated. Additionally, viscosity decreases at a higher rate at lower MgO contents, and after approximately 12 mol % MgO (8 mass % MgO), the curves tend to flatten out. This indicates a reduced rate of viscosity decrease at all temperature levels.

Regarding their impact on BF slag viscosity, it is judiciously suggested that CaO can be replaced by MgO on a molar basis. This is because both CaO and MgO are more effective at reducing viscosity at lower temperatures compared to higher temperatures. Such conclusions have also been supported by several researchers (Behera et al., 1990; Forsbacka et al., 2003; Iida et al., 2000; Machin and Yee, 1948).

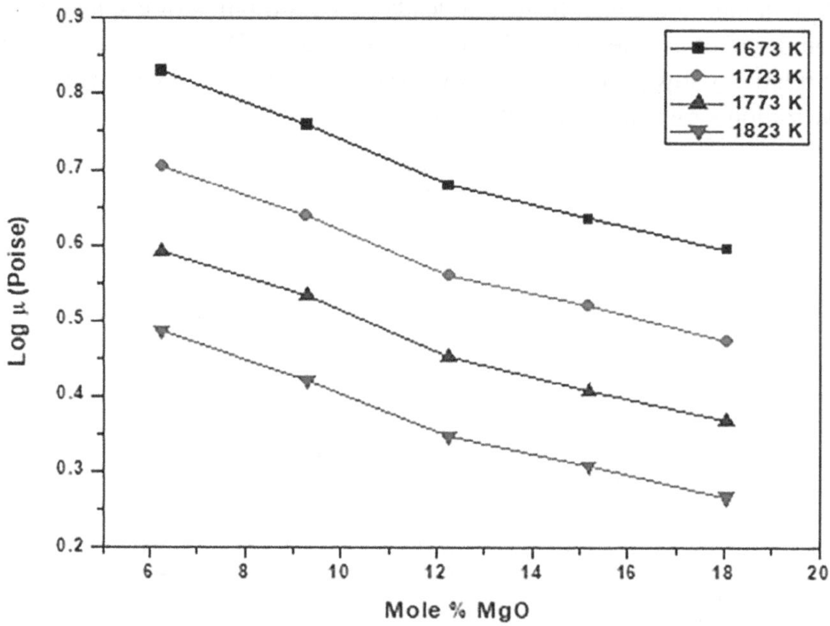

FIGURE 8.7 Log(μ) vs. mol % MgO at different temperatures.

E_μ values of the respective slags are calculated from the plots between $\ln(\mu)$ and $1/T$ (Figure 8.8). It is confirmed that viscosity variation with MgO content also follows the Arrhenius equation, and that the decrease in viscosity with increasing temperature results from the weakening of ionic bonds—not from polymerization or depolymerization of the silicate structure.

It can therefore be inferred that, at any specific MgO content, the degree of polymerization or depolymerization remains practically constant regardless of temperature variations.

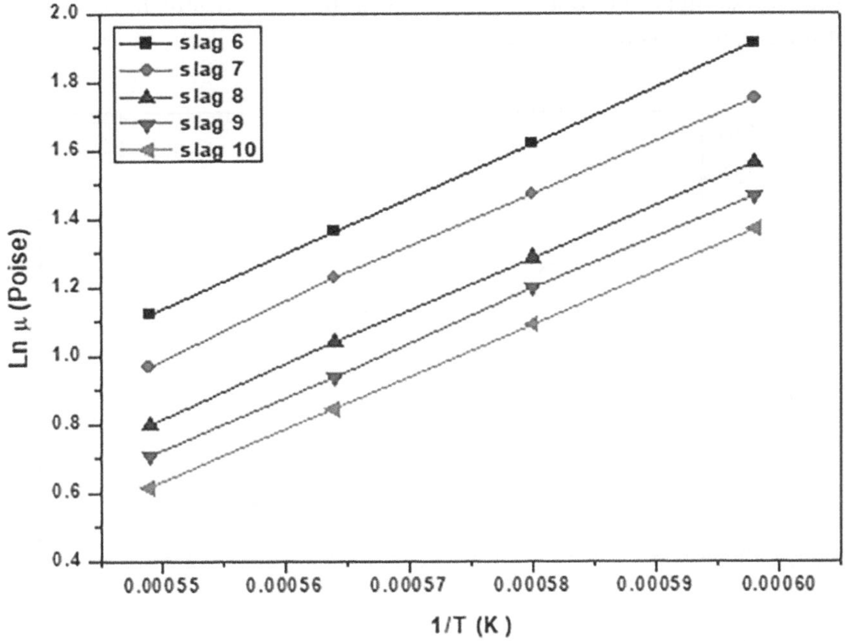

FIGURE 8.8 Variation of ln(μ) with 1/T.

The fluctuation in E_μ values computed from the plots in Figure 8.8 with variation in mol % MgO is shown in Figure 8.9. Up to approximately 13 mol % (8 mass %) MgO, E_μ declines significantly. Thereafter, it decreases at a comparatively slower rate until about 15 mol % (10 mass %) MgO. As the MgO mol % increases further, E_μ continues to decrease gradually, and the plot becomes nearly parallel to the X-axis.

This observation aligns with the findings of Singh (1989), who reported that for Indian BF slag containing 20 mass % Al_2O_3, MgO additions beyond 10 mass % have no significant effect on viscosity. Thus, it can be inferred that beyond 10 mass % MgO, additional content does not contribute meaningfully to the depolymerization of discrete anionic silicate structures.

A careful comparison between Figures 8.5 and 8.9 confirms that, with MgO variation, the corresponding E_μ values are somewhat higher than those observed for CaO variation. This suggests that MgO has a lower capacity than CaO to break down the silicate network into smaller flow units. This behavior is influenced by the values of Z/r^2, where Z is the cation valency and r is the ionic radius. Despite the larger Z/r^2 value for the

Mg^{2+} ion compared to that of the Ca^{2+} ion, the network-breaking capacity of magnesium oxide is lower.

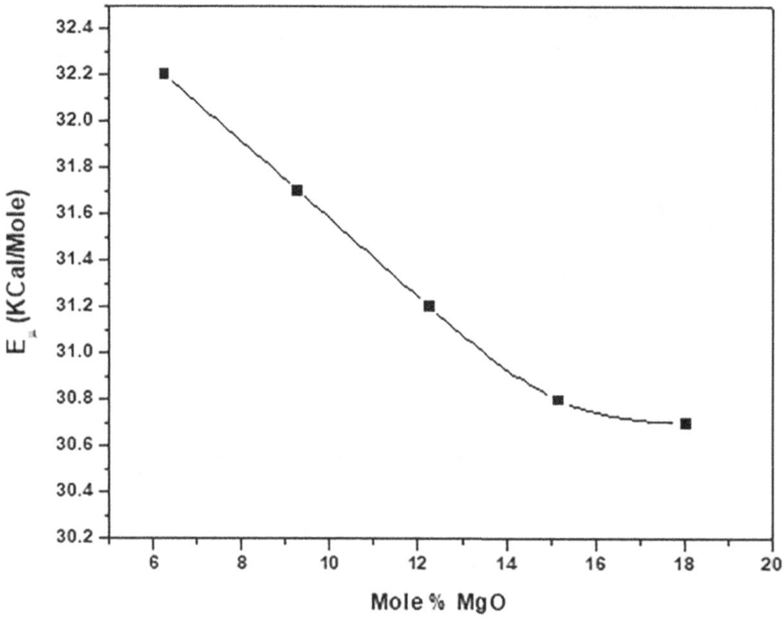

FIGURE 8.9 Variation of E_μ with mol % MgO.

Examining the relationship between viscosity and E_μ fluctuation with NBO/T is particularly fascinating. According to Mills (2011), NBO/T measures the degree of depolymerization by counting the number of non-bridging oxygen atoms per tetrahedrally coordinated atom. NBO/T is calculated using the following equation:

$$NBO/T = \frac{Y_{NB}}{X_T}$$

where;

$$Y_{NB} = \sum 2[X_{CaO} + X_{MgO} + X_{FeO} + X_{MnO} + 2X_{TiO_2} + X_{Na_2O} + X_{K_2O} - 2X_{Al_2O_3}]$$

$$X_T = \sum [X_{SiO_2} + 2X_{Al_2O_3}]$$

where; X is the mole fraction of the respective oxides in the corresponding slag.

Plots for log(μ) and NBO/T for each of the 10 slag samples considered for C/S ratio and MgO content variation, at a constant temperature of 1673 K, are shown in Figure 8.10. This plot illustrates the relationship between the degree of depolymerization and viscosity, accounting for changes in both the C/S ratio and MgO concentration. Figure 8.10 shows that viscosity is strongly correlated with the degree of depolymerization at 1673 K, with log(μ) following a declining trend as NBO/T increases.

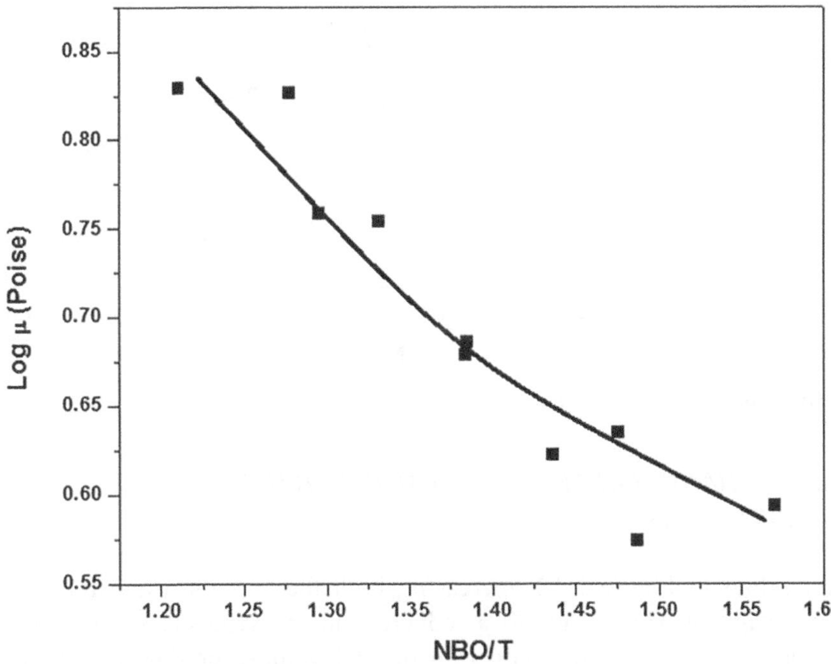

FIGURE 8.10 Log(μ) vs. NBO/T at 1673 K for slags 1–10.

The ability of CaO and MgO to break down the discrete anionic silicate network is not identical, as shown in Figure 8.11, which presents the variation of E_μ with NBO/T for all 10 slags. Here too, an increase in NBO/T generally corresponds to a decrease in E_μ, indicating that an increased degree of depolymerization—as represented by NBO/T—contributes to reduced activation energy for viscous flow.

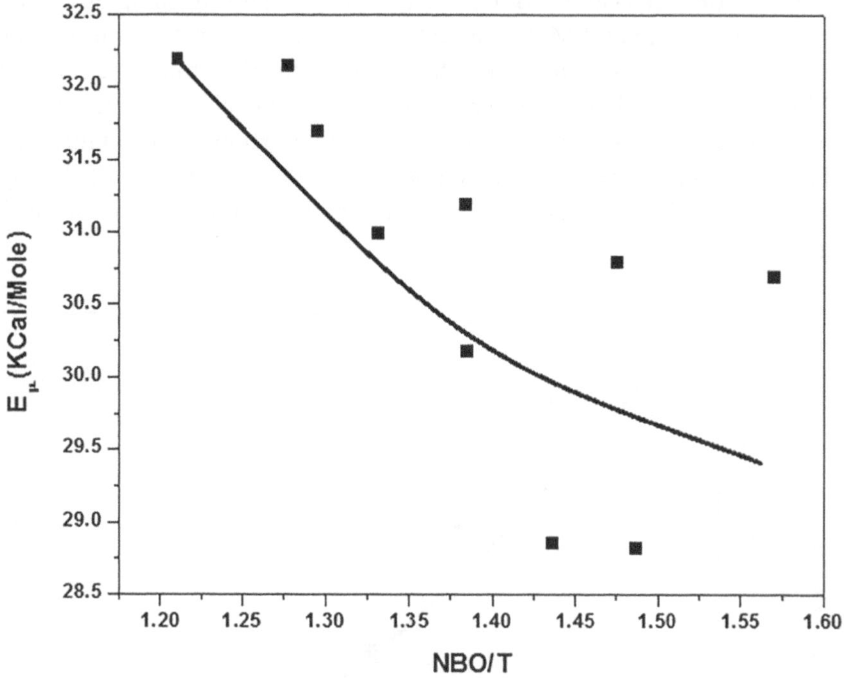

FIGURE 8.11 E_{μ} vs. NBO/T for slags 1–10.

8.3.3 CHANGES IN THE C/S RATIO AND MGO CONTENT SIMULTANEOUSLY

According to Figure 8.12, increasing both the C/S ratio and the mass percentage of MgO leads to a reduction in the viscosity of the corresponding slag, with this reduction occurring more gradually at higher values of both variables and at comparatively higher temperatures. Additionally, the viscosity of the slag decreases more sharply when the C/S ratio varies between 0.9 and 0.95, along with MgO fluctuations between 4 and 6 mass %.

Therefore, it is essential to carefully control the operating temperature when processing such slags in industrial conditions. If the temperature falls below a certain threshold, these slags may pose operational challenges, negatively affecting heat transfer in the furnace, slag-metal reactions and exchanges, and related processes. However, this issue is less likely to occur when both the C/S ratio and MgO content exceed 6 mass %.

Within the examined range of variables, the viscosity of the slag in such cases decreases more gradually with increasing temperature.

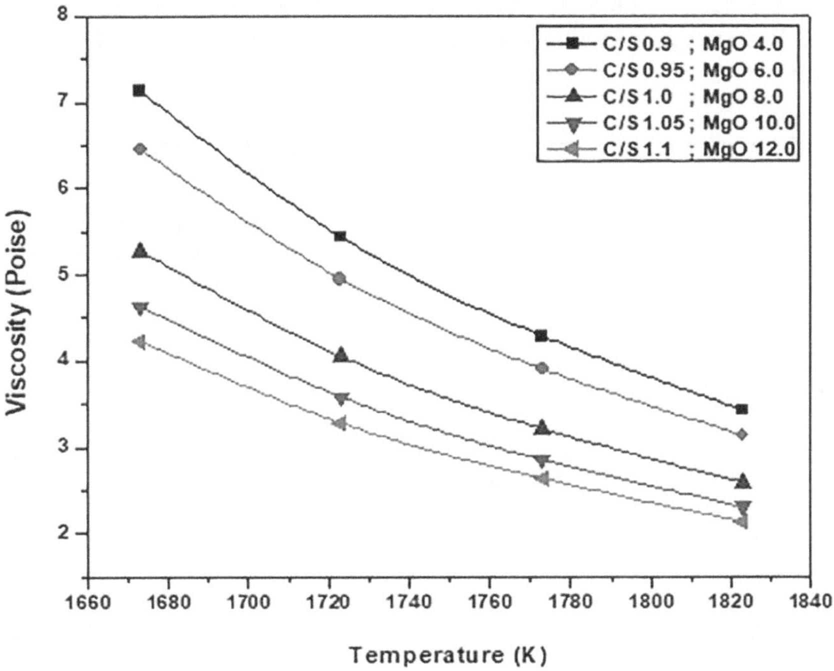

FIGURE 8.12 Viscosity vs. temperature of slags 11–15.

It is evident from Figure 8.13 that viscosity decreases with a simultaneous increase in both the C/S ratio and MgO mass %. However, for any given composition, the straight-line relationship is maintained even as the temperature increases. This finding confirms that temperature alone cannot cause depolymerization, although it may weaken ionic bonds and thereby reduce the E_{μ} value.

On the other hand, the observed decrease in E_{μ} with increasing C/S ratio and MgO content indicates a lower energy requirement to break the silicate network for viscous flow. It is important to note that cations with higher ion–oxygen interaction parameter (I) values—such as Al^{3+}, Si^{4+}, and Ti^{4+} (where $I > 1.16$)—are more capable of tightly binding oxygen ions within specific network structures, thus serving as network builders

Conversely, Ca^{2+} and Mg^{2+} ions, which exhibit lower I values ($I < 0.7$), have weaker ion–oxygen attraction and are therefore considered network breakers.

This demonstrates that the depolymerization process becomes easier and less energy intensive when slags contain larger amounts of CaO and MgO.

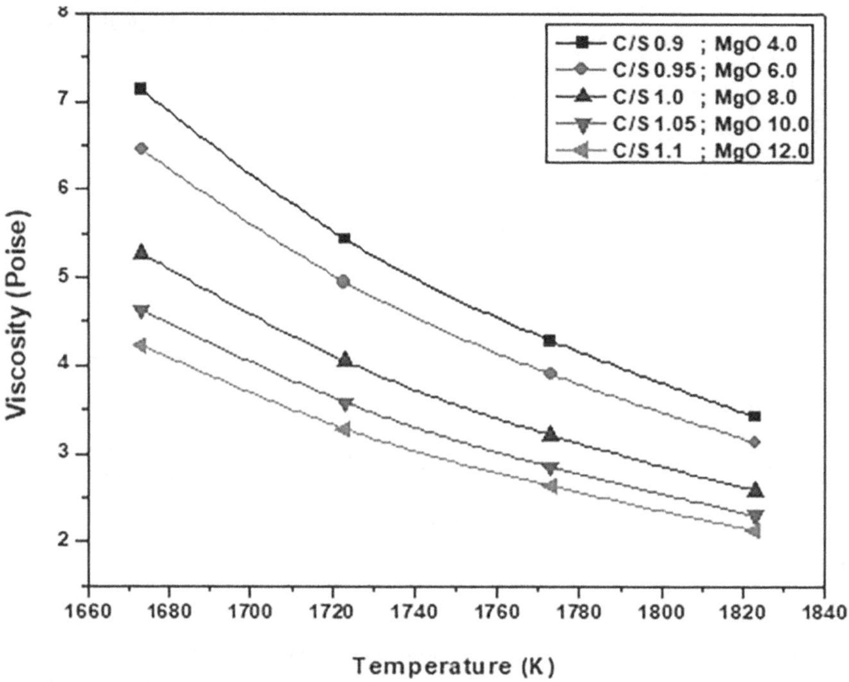

FIGURE 8.13 Ln(μ) vs. $1/T$ for slags 11–15.

The viscosity of slag decreases at a faster rate for lower basicity values, $B_2\left(\dfrac{CaO+MgO}{SiO_2}\right)$ and $B_3\left(\dfrac{CaO+MgO}{SiO_2+Al_2O_3}\right)$, while the rate of viscosity reduction slows as basicity increases (Figures 8.14 and 8.15). This behavior is due to the fact that as silicate flow units become smaller, they require proportionally more oxygen from alkaline earth oxides such as CaO and MgO. However, their gradual addition becomes less effective at further reducing the size of the flow units.

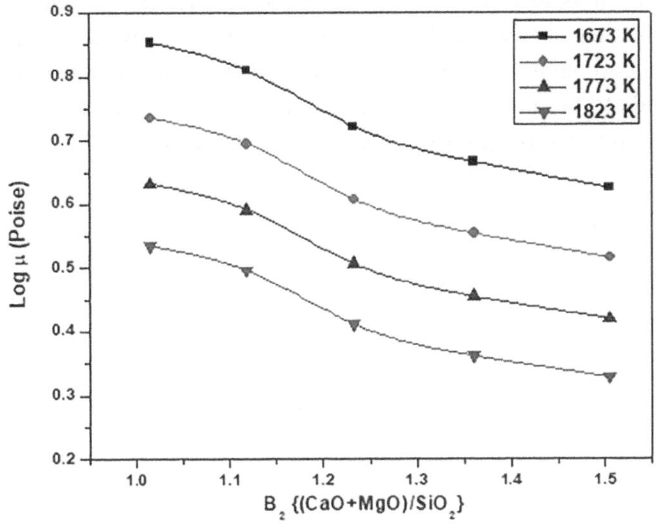

FIGURE 8.14 Variation of log(μ) with B_2.

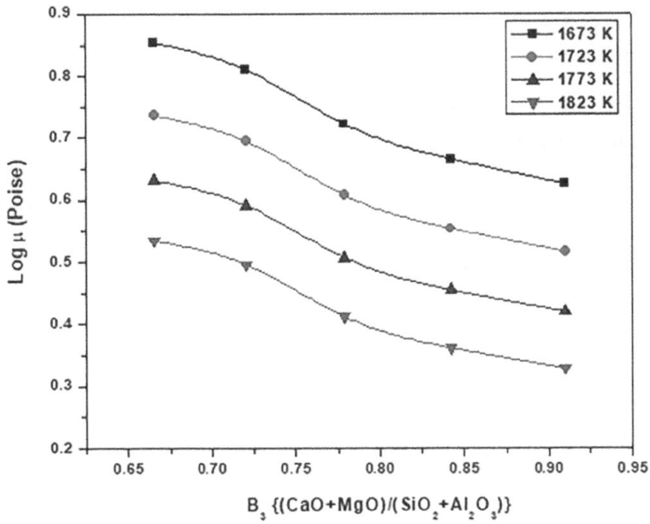

FIGURE 8.15 Variation of log(μ) with B_3.

The decrease in activation energy of viscous flow with basicities B_2 and B_3 is shown in Figures 8.16 and 8.17, respectively. In both cases, E_μ values continuously decrease, but at a slower rate as basicity increases.

This behavior is attributed to the reduced efficiency of oxide flux additions in further shrinking the size of the silicate flow units. As a result, the associated negative charge becomes less concentrated, requiring fewer ionic bonds to be broken or rearranged for the flow units to move relative to one another. Consequently, E_μ decreases more slowly.

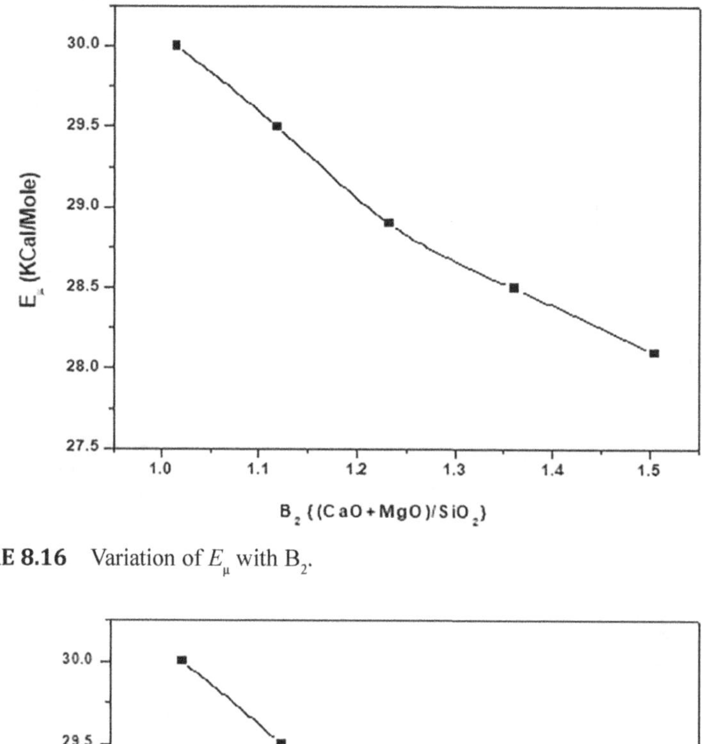

FIGURE 8.16 Variation of E_μ with B_2.

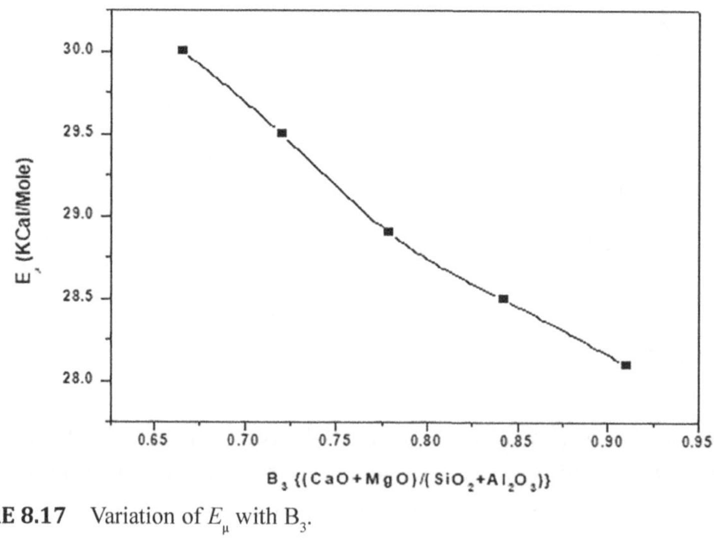

FIGURE 8.17 Variation of E_μ with B_3.

8.4 INFERENCE

- There is an inverse relationship between temperature rise and an increase in the C/S ratio in reducing slag viscosity, with temperature rise being more effective at lower C/S ratios.
- Similar findings apply to MgO variation; however, in general, MgO has a lower capacity to depolymerize the discrete silicate network compared to CaO.
- The level of depolymerization, as indicated by NBO/T, is responsible for the decrease in slag viscosity and the activation energy of viscous flow (E_μ).
- At MgO levels above 10 mass %, the material no longer participates in forming a stoichiometric slag structure.
- When working with slags having a C/S ratio below 0.95 and MgO content below 6 mass %, it is imperative to maintain stricter temperature control during industrial operations.

KEYWORDS

- **activation energy**
- **depolymerization**
- **industrial practices**
- **ionic bonds**
- **ion-oxygen interaction parameter**
- **network breakers**
- **pig iron**
- **stoichiometric slag structure**
- **viscosity**

REFERENCES

Behera, R. C., Mohanty, U. K., & Mohanty, A. K. (1990). Viscosity and constitution of slags associated with metallorthermic extraction of some less common metals. *High Temperature Materials and Processes*, *9*(2–4), 57–76.

Dash, S., Mohanty, N., Mohanty, U. K., & Sarkar, S. (2012). A novel technique for determination of flow characteristics of blast furnace slag.

Forsbacka, L., Holappa, L., Iida, T., Kita, Y., & Toda, Y. (2003). Experimental study of viscosities of selected $CaO–MgO–Al_2O_3–SiO_2$ slags and application of the Iida model. *Scandinavian Journal of Metallurgy*, *32*(5), 273–280.

Gupta, V. K., & Seshadri, V. (1973). Studies on viscosity of high alumina blast furnace slags. *Trans. Indian Inst. Metals*, *26*, 55–64.

Iida, T., Sakai, H., Kita, Y., & Shigeno, K. (2000). An equation for accurate prediction of the viscosities of blast furnace type slags from chemical composition. *ISIj International*, *40*(Suppl), S110–S114.

Machin, J. S., & Yee, T. B. (1948). Viscosity studies of system $CaO-MgO-Al_2O_3-SiO_2$: II, $CaO-Al_2O_3-SiO_2$. *Journal of the American Ceramic Society*, *31*(7), 200–204.

Machin, J. S., Yee, T. B., & Hanna, D. L. (1952). Viscosity studies of system CaO–MgO–$Al_2O_3–SiO_2$: III, 35, 45, and 50% SiO_2. *Journal of the American Ceramic Society*, *35*(12), 322–325.

Mills, K. (2011). The estimation of slag properties. *Southern African Pyrometallurgy*, *7*(3), 35–42.

Mohanty, U. K. (1998). *Thermo Physical Properties of Some Metallorthermic Slags* (Doctoral dissertation, PhD Dissertation, RE College, Rourkela).

Moore, H., & McMillan, P. W. (1956). A Study of Glasses Consisting of the Oxides of Elements of Low Atomic Weight: Part III. Certain Physical and Chemical Properties of the Experimental Glasses in Relation to the Glass Structure. *J. Soc. Glass Technol.*, *40*(193), 139.

Saito, N., Hori, N., Nakashima, K., & Mori, K. (2003). Viscosity of blast furnace type slags. *Metallurgical and Materials Transactions B*, *34*, 509–516.

Singh, B. N. (1989). Recent trends in iron and steelmaking technology. *Bulletin of Materials Science*, *12*, 435–447.

Singh, R. N. (1984). Viscosity measurements of high alumina blast furnace slags. *Steel India*, *7*(2), 73–83.

Sunahara, K., Nakano, K., Hoshi, M., Inada, T., Komatsu, S., & Yamamoto, T. (2008). Effect of high Al2O3 slag on the blast furnace operations. *ISIJ International*, *48*(4), 420–429.

Tiwary, J. N., Sarkar, S., Mishra, B., & Mohanty, U. K. (2013). Structural aspects of blast furnace slag. *Emerging Materials Research*, *2*(3), 152–162.

Weyl, W. A. (1951). The significance of the co-ordination requirements of the cations in the constitution of glass. III. The chemical durability of glasses. *J. Soc. Glass Technol.*, *35*(167), 462–468.

Index

G

H

I

J

K

L

M

Q

R

S

For Product Safety Concerns and Information please contact our
EU representative GPSR@taylorandfrancis.com Taylor & Francis
Verlag GmbH, Kaufingerstraße 24, 80331 München, Germany